THE LEAVING OF LIVERPOOL

A detailed record of the city's tramway abandonment programme

**Martin Jenkins
and Charles Roberts**

LRTA
Since 1937

The right to be identified as the Authors has been asserted in accordance with sections 77 and 78 of the Copyright Designs and Patent Acts 1988.

All rights reserved. No part of this publication may be reproduced, stored in a retrieval system or transmitted, in any form by any means, electronic, mechanical, photocopying, recording or otherwise, without prior permission in writing from the publisher.

Published by the Light Rail Transit Association
8 Berwick Place
Welwyn Garden City
AL7 4TU

www.lrta.org

Copyright © Light Rail Transit Association 2020

Designed by:
Steve Herbert-Mattick
sherbertdesign@gmail.com

Print and bound in the UK by:
Bishops Printers
Walton Road
Farlington
Portsmouth
PO6 1TR

ISBN 978-0-948106-60-6

Front cover:
Between 1931 and 1942, Liverpool Corporation built hundreds of high-capacity, double-decker trams at its impressive purpose-built works on Edge Lane. However, when the decision was taken in 1945 to scrap the 97 mile system, many were consigned to the scrap heap after a relatively short service life. Included were these two Marks Bogies seen approaching the peak hour terminal point at Old Haymarket in 1952. *Phil Tatt/Online Transport Archive*

Frontispiece:
Liverpool was renowned for its complex track work including sections of single track and loops and one-way workings. One short length on steeply-graded Beloe Street always attracted photographers. Opened as late as November 1935, the track formation had to fit within the existing pavement line. To alleviate the need for costly point work, there was a section of interlaced track as well as closely-spaced double track which prevented two cars from passing. Movements were controlled by electric signals. This view was taken on 8 September 1951, the final day of operation. *A S Clayton/Online Transport Archive*

Contents

Background and wartime ... 9

1945 ... 21

1946 ... 25

1947 ... 33

1948 ... 49

1949 ... 65

1950 ... 89

1951 ... 111

1952 ... 133

1953 ... 149

1954 ... 167

1955 ... 181

1956 ... 197

1957 ... 213

1958 and beyond ... 227

Introduction

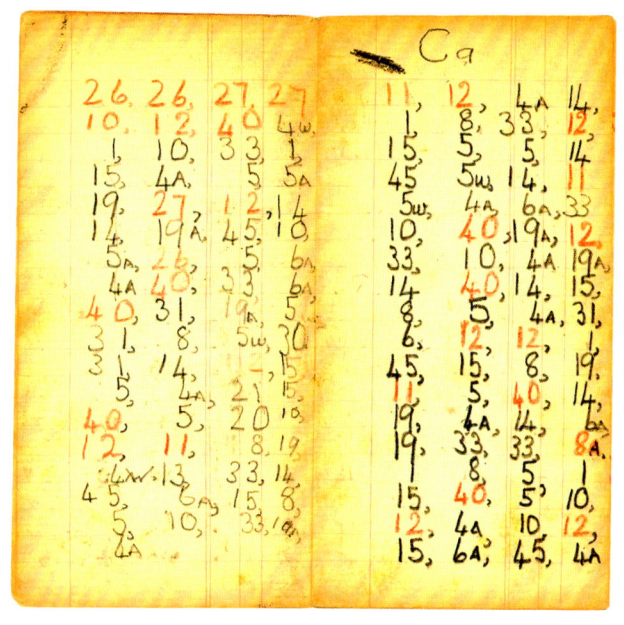

Few books have been written laying out in detail the run-down of a great British tram system. This is our attempt to do just that by bringing together all the known information although we are sure there is probably still more 'out there'.

There have been a number of memorable books devoted to the Liverpool tram system, including the magisterial five-volume *Liverpool Transport* by J B Horne and T B Maund and the very personal, highly-evocative *Edge Lane Roundabout* by Brian Martin. There have also been invaluable personal recollections recorded over the years in the newsletters issued by the Merseyside Tramway Preservation Society. This volume seeks to complement these, and others listed in the bibliography, by following, in diary form, the detailed decline of one of the finest city tramways in the country.

Despite investing in hundreds of new trams and opening an extension as late as 1944, the City Council approved a complete conversion programme the following year. What led to this sudden change of policy? How far were powerful, local politicians involved? Were the figures produced by The Transport Department an entirely accurate representation of the post-war alternatives? Did they include the cost of track removal and road resurfacing? Had trams been built, and extensions laid, to the highest standard? Was there a breakdown in communication between different areas of the Transport Department? Did wartime neglect and post-war shortages lead to the inevitable? Why was so much track relaid after the war only to last a few years? Could/should any part of the network, with its miles of segregated 'grass tracks' have been retained?

Personal insights

Martin: "My fascination with Liverpool trams began at an early age. My earliest memories date from the war when we made periodic visits from Rhyl where my mother and I were evacuated from our home in Wallasey. Later, I began to ride the tram routes on a Saturday whilst my mother spent the morning shopping. Our favourite spot for lunch was the Kardomah café on Church Street where, armed with different coloured pencils, I aimed for a window seat so I could record the routes passing beneath. Later I graduated to logging fleet numbers.

What I didn't realise was that the trams were on the way out. They just seemed so permanent. As the years progressed, I joined the LRTL and made many friends, all of whom shared my passion for 'our trams'. These included Brian and Jeff Martin, Ronnie Stephens, Tony Gahan, Don Littler, Jerome McWatt, John Horne, Chris Bennett, Tony Packwood, Nigel Eames, Terry Martin, Dave Webster, Alf Jacob, Leo Quinn, Freddie Lloyd-Jones, Bob Prescott, John Maher, Cliff Noon, Eric Pollard, Bert Roberts, Norman 'Nobby' Forbes, Jack Gahan, Ted Gahan, Edward Piercey, Stan Watkins, Harry Dibdin, Ted Fowler and Messrs Dale, Hill and Higgins. Sadly, all too many are now no longer with us.

Gradually, I started building up a collection of photographs with the view to obtaining a shot of each and every car, an objective shared with Brian Martin (overleaf, top left, with younger brother Jeff). Right up until his untimely death in 2001 we would regularly exchange images. Top of our 'wants' lists were those cars destroyed in the disastrous fire at Green Lane depot in November 1947, especially Baby Grand 300, as well as those with very short lives such as Baby Grands 217, 225, 228 and 259. I remember Brian was thrilled when a view of Liner 888 came to light just before he died. We also wanted tangible reminders as well. This led to the acquisition of parts of trams, stop signs and all manner of paperwork.

Then in 1959/60, I was privileged to be the first chairman of the Liverpool University Public Transport Society and the Merseyside Tramway Preservation Society, which eventually led to the preservation of Streamliner 869 of 1936, now a prize exhibit at the National Tramway Museum. Subsequently Baby Grand 245 and English Electric bogie car 762 have been restored. In the ensuing 60 years, I have never lost my fascination and it has been exhilarating collaborating with John Horne, Geoff Price and Charles Roberts in order to produce as accurate an account as possible. Along the way we have made a few important factual discoveries and solved some 'mysteries' although a few remain. All we know is now recorded for posterity."

Charles (above right): "I need to start by saying that I have no first hand memories of Liverpool's trams at all, not being born until two years after they had been abandoned. Although I was born and brought up in Wirral, both sides of the family had lived in Liverpool until the 1940s, so hearing stories about 'going into town on the tram' were very much part of my upbringing. I can vividly remember being loaned a copy of one of the MTPS booklets published in the 1970s and been astounded to see pictures of modern-looking trams sweeping along grassed tram tracks and separated from road traffic by neat privet hedges. Where had it all gone? A bit of background reading quickly gave me the answer and I was able to look at Liverpool's roads after that and try and imagine what they had been like in my parents' and grandparents' days.

Although I never got actively involved in the tram preservation scene, I took great interest as, one by one, the surviving Liverpool cars were returned to active service. And helping Martin with this book, my eyes have been further opened to some of the remains of the system which this volume describes. In looking for rosettes on buildings, I made one personal discovery that brings the story full circle. There is still a rosette on 255 Breck Road in Everton, nowadays a barber's shop. In the 1891 Census (pre-tram days, of course), it was listed a butcher's shop. The butcher was Ernest Maguire my great-grandfather."

Acknowledgements

We are especially indebted to John Horne and Geoff Price, each of whom has a life-long interest in Liverpool trams. Both have made their extensive archives available and also offered constructive comments and observations. John is seen on board Baby Grand 245 in 1957 whilst Geoff is at the controls of ex-Double Staircase car 601 at Prince Alfred Road depot in 1946.

We are equally indebted to those photographers, past and present, who recorded precious images of the trams especially local schoolmaster and LRTL representative, Norman Forbes (opposite, top left), without whose photographs many parts of the north end of the network would have gone unrecorded. He is seen in this view at Bowring Park in 1955.

Other local photographers whose images we have used include: Jack, Ted and Tony Gahan, Brian Martin, John McCann, Cliff Noon, Bert Roberts, Leo Quinn and Stan Watkins. Visitors Roy Brook, Freddie Lloyd-Jones, David Packer, Bob Parr, Henry Priestley and Dick Wiseman took many outstanding images and

Swallow and Dave Thomas for some insights into the workings of the Transport Department. We are especially grateful to the Light Rail Transit Association and Carl Isgar for supporting this project.

Maps and plans

It is probably true to say that all published maps of the Liverpool tram system derive from one 'master', meticulously prepared by John Horne, with assistance from John Maher, in about 1960, copies of which were sold in support of local tramway preservation. We have based ours on the version which appears in *Liverpool Transport, Volume 4*, with permission from Venture Publications (the successors to TPC). Our city centre enlargement is based on a version created by R A Smith for the LRTA in *The Tramways of South Lancashire and North Wales* and is again used with full permission. Annotations have been made by the authors, who accept responsibility for any errors which may have crept in. Depot diagrams have been created by Chris Poole from contemporary OS Maps. These include specific track details unearthed by Geoff Price in relation to the layouts at two of the south end depots.

Featured images and documents

It must be borne in mind that for much of the period covered by this volume, the cameras used by many enthusiasts were quite primitive by today's standards and most of the film stock required 'perfect' weather conditions for the best results. Although the authors have had access to original negatives in some cases, the source material has been variable. Modern digital enhancement techniques are able to get the best out of some less-than-perfect pictures, but a few which do not reach this standard have been included because of the unique nature of the subject matter.

The decision was made to base the bulk of the story around black and white coverage. Although colour photography was becoming more widespread in the 1950s, the vast majority of colour images of the Liverpool system relate to the final few months, and many of the best images have been published before, often multiple times, whereas a significant number of the monochrome images in this book have not.

Photographs are supplemented by contemporary leaflets, notices, tickets, time-table extracts etc. Many of these have a careworn feel to them – time-tables were well-thumbed, notices were salvaged from cars after the abandonment date had passed (occasionally just before!), tickets have annotated notes scrawled on them and so on. We feel that authenticity is achieved by presenting these 'as is' rather than by artificial enhancement.

And finally, we have tried to use forms of wording which were in use during our period, so although it might seem strange to read the word time-table with a hyphen today, that is how it appeared in official documents. The Corporation could, however, be very inconsistent with place names – something we have made reference to in subsequent chapters – but we have endeavoured to use the most common forms in our own writing, such as Pagemoss and Five Ways.

Glossary

Much use has been made of quotes from written journals and correspondence, and from other contemporary notes. To save space, the attribution of many of these is through individuals' initials: Tony Gahan (AFG), Ted Gahan (EAG), Geoff Price (GWP), John Horne (JBH), Jack Gahan (JWG), Martin Jenkins (MJ).

their work is also featured. Despite efforts to establish their origin, a few images remain unattributable, so they are credited to the collection from which we obtained them.

The authors have been fortunate to be able to draw on the detailed notebooks and diaries kept by the three remarkable Gahan brothers, each of whom possessed an encyclopaedic knowledge of the Liverpool system. The view below shows the brothers at the Wirral Transport Museum in December 2006. From left: Jack (1921-2009), Ted (1926-2016) and Tony (1940-2019).

We should also like to thank the National Tramway Museum (Laura Waters) and the Merseyside Tramway Preservation Society (David Taylor) for access to their collections; Hugh Taylor for access to the Peter Mitchell collection; Neal Watkins for access to the Stan Watkins collection; Bob Bradshaw for access to the collection of tickets amassed by Bob Prescott; and Ken

Online Transport Archive

As with the authors' other transport-related books, this volume has been compiled in conjunction with Online Transport Archive, a UK based charity dedicated to the conservation and preservation of transport images and to which the authors' fees have been donated.

For further detailed information about the archive, please contact the Secretary at 25 Monkmoor Road, Shrewsbury SY2 5AG (email secretary@onlinetransportarchive.org).

Bibliography

Much of the material in this book has come from primary, contemporary sources, supplemented by extracts from time-tables, pamphlets, newspapers, record books and magazines, as well as a number of publications, most notably:

A Nostalgic Look at Liverpool Trams 1945-1957, Steve Palmer and Brian Martin, Silver Link Publishing, 1996
Busman, Bill Peters, DTS Publishing, 2002
By Tram to Garston, Eric Vaughan, Merseyside Tramway Preservation Society, 1986
Edge Lane Roundabout, Brian Martin, Merseyside Tramway Preservation Society, 1984
Green Goddesses Go East, Ian L Cormack, Scottish Tramway Museum Society, 1961
Liverpool Corporation Tramways 1937-1957 (three volumes), T J Martin, Merseyside Tramway Preservation Society, 1972-1995
Liverpool Corporation Transport, Fleet History PC13, PSV Circle/Omnibus Society, 1979
Liverpool Trams Fleet List, T J Martin, Merseyside Tramway Preservation Society, 1978
Liverpool Tramways (three volumes), Brian Martin, Middleton Press, 1997-2000
Liverpool Tramways 1943 to 1957, R E Blackburn, LRTL, c1970
Liverpool Transport (volumes 3-5), J B Horne and T B Maund, Transport Publishing Company, 1987-1991
The Tramways of South Lancashire and North Wales, J C Gillham and R J S Wiseman, LRTA, 2003

Martin Jenkins
Walton-on-Thames

Charles Roberts
Upton, Wirral

January 2021

CHAPTER 1 | Background and wartime

This book is concerned with the post-war run down of Liverpool's trams and why it was decided to abandon the 97 mile system with it miles of reserved track and hundreds of modern trams. It is also concerned with the often fractious relations between the various General Managers and the powerful Chairmen of the Transport Committee; between 'Hatton Garden' (location of the Transport Department Head Office) and the powers-that-be at Edge Lane Works; between the works and the individual depots who resented instructions from above; between Hatton Garden and the City Engineers & Surveyors Department (CE&S) who were responsible for anything relating to the track; and between management and the workforce. On top of this, the Council pursued a policy of employing local labour rather than using, often far cheaper, outside contractors.

It all began in the middle years of the nineteenth century when horse buses began to serve the city's burgeoning population which had increased hugely from 78,000 in 1801 to 684,958 in 1901. Although a street tramway operated briefly in Old Swan in 1861, it was not until 1869 that a start was made on developing a network of horse car routes. Owned latterly by the Liverpool United Tramways & Omnibus Company, the entire system was sold to the Corporation for over half a million pounds in 1897.

Electrification followed rapidly, with the first single-decker trams being imported from Germany and the USA. By 1901, everything had been electrified except for a solitary horse car line in Litherland which survived until 1903. The system expanded rapidly with many cars being built by the Corporation at their Lambeth Road works with Bellamy top covers (named after the General Manager of the time) which were also fitted to former open-top cars. Subsequent new cars also carried the names of later Managers including, for example, Mallins Balconies and Priestly Standards. The livery was crimson lake and ivory although some cars were painted overall white for a first-class service operated between 1908 and 1923.

In September 1914, Liverpool led the way by opening its first section of central reservation built to serve new 'garden suburbs' and relieve the severe overcrowding in the older, poorer parts of the city. This was also the year in which route numbers were introduced.

In 1928, a massive new works plus a separate adjacent depot was opened on Edge Lane. Under Percy Priestly's management, more Standards were built to improved specifications aimed at improving riding quality, overall speed and reducing operational costs. Some were also totally enclosed. In 1929, an experimental single-decker No 757 was produced but it was unsuccessful and scrapped. Next came 12 capacious, high-capacity eight-wheel cars. Although conventional in design, these Priestly bogies proved popular and changed public perception of the tram. However, throughout this period, Priestly was under severe pressure. Many people disliked the older trams finding them slow and uncomfortable. These concerns were shared by local councillors, some of whom were on the powerful Transport Committee. Calls were growing for the trams to go. By now, the population had increased to 855,688. However, there were few labour-intensive factories offering mass employment so many found work in the docks and dock-related industries with many relying on the trams.

National Tramway Museum

When Walter Grey Marks became General Manager of the fourth largest municipal transport undertaking in the country in 1934, it was assumed he had been appointed to oversee a conversion programme much as he done at his previous post in Nottingham. However, after careful consideration, Marks successfully argued the financial case for retention, extension and modernisation. He succeeded in restoring the tramway's profitability and improving relations with the Transport Committee, although some members were still anti-tram. As with his predecessors, problems relating to the track and surrounding road surface continued to cause problems. Retrospectively, it is clear that far too little had been invested in new rail so that by 1939, much of the track was in urgent need of renewal. Furthermore, the city's renowned reserved tracks had not been laid to the highest standard.

Relations with the neighbouring Borough of Bootle were also complicated so it was expected that when Liverpool's operating lease expired in 1942 the routes within the Bootle boundaries, including those serving Seaforth (seen here), would be abandoned. Despite this threat, Litherland depot was actually modernised. A spacious but shallow extension housing 40 cars on eight roads was built on the north side of the older eight-road structure. New offices were also provided. When opened on 29 September 1939, the depot now housed 80 cars on 16 roads.
[Above] A G Wells

Photographer unknown

Marks quickly made changes to the design of the high-capacity Cabin cars currently being built at Edge Lane. His aim was to reduce power consumption and track damage by reducing overall weight. Unfortunately, this new generation of trams suffered from structural, mechanical and electrical problems with heavy rain, snow and slush flooding motors and resistances. In 1936, Marks released a damming internal report on labour relations at the works. He highlighted ingrained problems with poor equipment, inefficiency, over-manning, shoddy work, 'Spanish practices', lack of skills, no organised or planned maintenance schedule, inferior management and outdated, poorly maintained equipment. As the war progressed, these difficulties intensified with skilled men enlisting and those that remained being engaged in war work such as assembling jeeps and tank carriers. All this occurred as demand for public transport escalated dramatically. On the credit side, surviving staff at Edge Lane worked weekends and nights trying to keep the neglected fleet on the move.
[Top right] F N T Lloyd-Jones/Online Transport Archive

Although committed to tramway retention and expansion on selected corridors, Marks was a pragmatist and had identified sections with life-expired track for early replacement. If the war had not intervened, West Derby route 12 would have been curtailed at Queens Drive and 'Belt' routes 26 and 27 with their awkward track layouts and complex junctions replaced by buses. In the event, only the outer section of route 15 was abandoned when it was cut back from Croxteth Road to Princes Park Gates in May 1938.

This next section deals with the war years from which the tram system emerged in a parlous state leading to renewed calls for abandonment. It also illustrates the extraordinary burden placed on the increasingly neglected trams and track as the management tried to maintain a sufficient level of service with evermore people using public transport.

1939-1945: some key dates

1 September 1939: There were now 5288 staff and manual employees in the transport department. Blackout regulations led to the trams being fitted with headlamp masks, anti-blast netting on the windows, blacked out interior lights, blue painted bulbs and white edges to fenders and steps to improve visibility. A device was also fitted to the overhead to prevent arcing from section feeder joints at night. All illuminated stop signs, signal lights and bollards were disconnected with fog batons being used on single track sections. By May 1942, there had been a 1281 blackout incidents involving trams; for example, sometime in 1940, English Electric bogie car 762 collided with Priestly Standard 675 at South Castle Street. *[Opposite, top left] Leo Quinn collection/ Online Transport Archive*

3 September 1939: War declared on Germany. Tram scrapping stops and scrap parts retained as spares.

11 September 1939: Route 15 reinstated to Croxteth Road although no work was done on the track and the former trolley reverser was not reinstated.

18 December 1939: Crews issued with a revised Rule Book entitled 'Procedure in Event of an Air Raid'. Should one occur, they must immediately leave their tram and place a cloth over the headlight. To cope with the aftermath of any gas attack, former passenger cars 306, 507, 543, 550, 558, 561, 564, 566, 575 were designated as Air Raid Precaution (ARP) cars in which clothes could be decontaminated. When this threat receded,

10 | THE LEAVING OF LIVERPOOL

some became snowploughs in late 1942. ARP training had started in 1938 and air-raid shelters were built at Hatton Garden, Edge Lane Works and all depots.

February 1940: 75,000 an hour carried at peak times.

12 March 1940: Knotty Ash spur reactivated. Roy Thomson remembers "Early part of WW2 I was travelling from the City to Finch Lane. I thought any car on route 10 would suit me. One came along which I think was shown as 10A but did not realise any car terminated before Pagemoss. I was amazed to find myself being diverted from the main route. By this time I was the only passenger. I wondered why the crew looked rather strangely at me. I was only 11 years of age at the time and perhaps rather dumb!" *[Below] Leo Quinn collection/Online Transport Archive*

15 April 1940: Fare stages shortened due to rising costs and 7d all day ticket abolished, never to be reinstated.

June 1940: As more men joined the military, the first of an eventual 1200 women conductors were employed. The Corporation also made use of voluntary conductors. These civilians helped control movements on and off cars so that conductors could collect fares. A rise in absenteeism during the war years led to frequent crew shortages.

4 June 1940: Capacity at Garston depot increased when three tracks, with pits, were opened in the new bus garage.

15 July 1940: Routes 13, 19 and 44 extended to Gillmoss Lane along the first part of a new reserved track extension built to serve Napier's aircraft engine factory at Gillmoss and a new Royal Ordnance Factory (ROF) at Kirkby.

Autumn 1940: First of a series of supplementary queue points used by scores of peak hour extras were established. These eventually existed at Commutation Row, Roe Street, Old Haymarket, Victoria Street and North John Street. It meant fewer cars using the Pier Head and were generally in force from 4pm to 6.30pm Monday to Friday and noon until 6pm on Saturdays.

August 1940 to January 1942: Merseyside suffered heavily during the early part of the war. The Luftwaffe targeted both sides of the river in series of brutal air-attacks aimed at crippling the dock system. Just under 70 air-raids, the worst in May 1941, resulted in 1453 deaths in Liverpool and 257 in Bootle with many more injured. 7610 houses were destroyed and 120,000 damaged in Liverpool and Bootle, rendering thousands homeless. Despite the severity of the attacks, the docks continued but, at times, the trams ceased altogether. Following most raids there were curtailments and diversions but routes 7, 30 and 31 were not reinstated after the May 1941 'blitz'. Track was damaged 75 times and the overhead in excess of 250 times with repair gangs coming from Manchester and St Helens to assist. Several depots were hit together with the two power stations providing the current. Amazingly, only five trams were destroyed or damaged beyond repair: 31, 37, 228, 773 and 786. Another 232 suffered blast damage but were repaired. The first view shows the roof of 773 lying in the road and the second the broken remains of 37.

[Previous page and above] Brian Martin/MTPS; Martin Jenkins collection/Online Transport Archive

Priestly Standard 118 was one of many to suffer less severe blast damage. Note the 'red-on-white' 13 and 14 on the screen. This colour combination indicated a car was either terminating at South Castle Street or short of the Pier Head.

[Bottom left] Martin Jenkins collection/Online Transport Archive

28 October 1940: Route 19 extended to Stonebridge Lane. To handle the shift-change crowds and avoid delays on the main line, work began on constructing a half-mile double-track siding serving the Napier factory. Complete with crossovers and designated loading points, it was first used in 1941 together with a much shorter siding at Stopgate Lane on Walton Hall Avenue.

3 December 1940: The roof was ripped off new Baby Grand 297 when it derailed and overturned at the junction of Grove Street and Upper Parliament Street. Eight people injured. Casualties would have been far worse had the windows not been covered with anti-blast netting. *[Below] Leo Quinn collection/Online Transport Archive*

12 | THE LEAVING OF LIVERPOOL

1940: During the year, Bellamy roof cars 539, 540, 553 and 555 were withdrawn and placed in store.

1940/41: This particularly vicious winter exposed some of the limitations of the modern trams. They failed to function in heavy snow or torrential rain as their motors, cabling and resistances were too low and quickly became soaked or caused shorts. One remedy was to cover the underneath vent and for staff to light braziers in the depot pits in an attempt to dry out the motors. However, if the insulation had blown then the car had to go to Edge Lane where it could gather dust for weeks.

In the winter, the CE&S was responsible for keeping the streets clear of snow and the Transport Department the reserved tracks but neither had any snow-clearing cars. Under-powered Bellamys, fitted with two planks in the shape of a vee, were no match for deeply-packed snow. Eric Vaughan recounts in his book *By Tram to Garston* "the gangers had their heads and shoulders covered with sacking and one had sacking wrapped round his feet".

Working conditions in the winter were so bad, platform staff demanded that vestibules should be fitted to the remaining open-fronted cars so by August 1944, a robust 'Austerity' screen with a sloping glass panel at the front had been fitted to Nos 9, 46, 58, 68, 80, 106, 131, 142, 143, 322, 330, 333, 376, 378, 450, 599, 608, 636, 637, 640, 641, 644, 651, 656, 659, 661, 666, 678, 684, 692, 703, 707, 708, 718, 719 and 739. Subsequently, similar screens were fitted to 23, 29, 30, 50, 119, 133, 358, 368, 687, 736 and 741.

[Above] Martin Jenkins collection/Online Transport Archive

1941: Despite severe shortages of new rail, the CE&S used available stock from its yard at Breckside Park to relay some 4½ miles of track during each year of the war.

At some unknown date, dockers' service 36 (Gillmoss to Seaforth) was rerouted via Everton Valley and Kirkdale Vale in lieu of Spellow Lane which meant cars had to reverse twice on the morning runs.

25 July 1941: An unidentified Streamliner on route 10 jumped the rails in Fairfield after colliding with a lorry, stopping within inches of a shop window. *[Above] Leo Quinn collection/Online Transport Archive*

August 1941: Peak hour services 37/38 (Old Haymarket to Penny Lane) discontinued and replaced by additional workings on routes 4 and 5. As a result, older cars such as 303, 544, 614 and 618 probably placed in store.

19 January 1942: 230 trams out of action due to heavy snow. Motors and wiring damaged by excessive use of salt aimed at keeping track and pointwork clear. Many also had shattered lifeguards and trays.

3 February 1942: A fire at Green Lane depot destroyed cars 19, 171, 337, 671 and 989. Councillor Bessie Braddock asked why it took 20 minutes to access water supplies. Despite the fear of air-raids, only the new part of Litherland depot had sprinklers and no depot had a direct line to the Fire Brigade.

[Below] Leo Quinn collection/Online Transport Archive

5 February 1942: 116 Liners and Baby Grands inactive due to heavy rain affecting their wiring. Faced by a chronic shortage, among cars returned to front-line duty were Bellamy roof car 544 of 1910 and English Electric Balcony cars 614 and 618 of 1919.

Background and wartime | 13

9 February 1942: Another fire at Green Lane! Fortunately no losses. Some cars manhandled onto the street. All too often cars with electrical faults left overnight with poles on the wire.

March 1942: Routes 13, 19 and 44 extended on central reservation to the City boundary at Gillmoss Brook. To overcome wartime shortages, lighter rail was used, concrete instead of wooden sleepers, second-hand poles from Coventry, and shale and crushed brick for ballast to prevent the 'grass tracks' becoming mudbaths. As each new section opened, the connecting buses to Kirkby were curtailed accordingly. As the long extension to the ROF at Kirkby was deemed essential to the war-effort, it was constructed under a 1942 Order granted under the Emergency Powers (Defence) Act of 1939. It was therefore owned by the Government but maintained and operated by Liverpool Corporation who had the option to purchase the line at the end of hostilities. Many employees, including men engaged on vital maintenance at Edge Lane, were involved in its construction. This 1952 view captures its semi-rural backdrops. *[Above] Phil Tatt/Online Transport Archive*

April 1942: To save on wear and tear, many stops were eliminated and orderly queuing introduced with staggered stops especially in the city centre.

30 June 1942: The lease enabling Liverpool to operate the trams within the Borough of Bootle was renewed for the duration but only an annual basis. Bootle Corporation openly favoured buses and made little or no effort to maintain its already fragile track and overhead.

1 July 1942: Cheapest adult fare increased to 1½d. With fewer cars, longer queues and irregular time-keeping, passengers were encouraged to use 3d and 4d transfer tickets to reach their destination by taking the first car and changing at recognised transfer points.

July 1942: The unnumbered dockers' route from Seaforth to Pagemoss extended to Long View Lane, cars showing 10C in one direction and 18 in the other.

September 1942: After suffering fire damage on Townsend Lane, 225 was written off. This rare photo shows the car following an earlier accident. It had only entered service in April 1938. *[Opposite, top left] Martin Jenkins collection/Online Transport Archive*

17 October 1942: The last new tram entered service. No confirmed photograph of 300 has ever come to light. "A notable feature was the front panel (don't know whether A or B end – the A end was where the main trolley cable from the roof was located) was a lighter shade of green than the rest of the car." (Anthony Henry) Like 299, it differed from other Baby Grands by having an unusual upper deck window arrangement. The former is seen at the Pier Head, the city's river front terminal in April 1946. Either this or 300 rode on the truck formerly under 228.
[Opposite, left] Martin Jenkins collection/Online Transport Archive

1942: At some unknown date, a lengthy siding was built at Southbank Road on Edge Lane. This prevented the mainline from being obstructed during peak loading times. Cars usually showed 'Edge Lane/Southbank Road' on their destination screens.

unknown date, by passenger cars 55 and 65. It is understood the blades were removed when a passenger car returned to traffic. Bellamy 506, which had lost its top deck in the Green Lane fire returned to this depot as a single decker salt car. In 1947, 565 is seen outside Dingle depot. It was the only Bellamy to be painted green. Dingle had been extended in January 1938 when a new two-bay 12-track extension was built on the east side of the original 1899 building. The extension was approached via two full-width entrances, the eastern one giving access to the older building. Total capacity was now 140. *[Below] A C Noon/Online Transport Archive*

1 March 1943: Catering for the growing number of men who lived in the southern suburbs but were employed in and around the North Docks, an unnumbered workmen's service was introduced between Penny Lane and Seaforth with cars working over parts of the 46 and 18A.

Following an aerial attack on a Corporation bus, the white roofs on the Liners and Baby Grands were covered in a drab brown 'wash' to make them less visible from the air. On some, layers of cheap varnish or blueish grey paint was applied to the body work.

December 1942: To avoid major disruption caused by frozen points and packed snow on the reserved tracks, steel blades were fitted to cars 122, 134, 303, 326, 513, 539, 540, 546, 553, 555 and 565 which were then assigned to various depots. Ex-Double Staircase 597 was also equipped, as were passenger cars 18 and 717. Then, in early 1943, 502 and passenger cars 302, 376, 639 and 654 joined the snowplough fleet followed, at some

Background and wartime | 15

29 May 1943: During the early evening, Marks Bogie 854 lost air pressure descending the grade on Upper Warwick Street. As the driver was unable to stop the heavy car on the handbrake it hit the curve into Park Place at 30mph, toppled over into an empty shop and crushed a 14 year old girl who was on the pavement. The top deck was cut away to free those inside. Most of the 36 injured were cut by flying glass as the trams were not fully fitted with safety glass. *[Above] Martin Jenkins collection/Online Transport Archive*

23 June 1943: Complying with the Formation of Queues Order of 1942, the first of a number of tubular steel 'queue covers' appeared on Lime Street. These primitive structures offered little protection from the city's infamous 'slantindicular' rain but they did force passengers into orderly lines. Early examples were constructed from recycled wood and corrugated iron used to protect goods arriving from the USA. More closely-spaced stops were eliminated in an attempt to reduce wear and tear on braking equipment, a move repeated in December.

14 September 1943: Route 29 extended on roadside reservation from Muirhead Avenue East to Lower Lane via Dwerryhouse Lane and Lowerhouse Lane (sometimes Lower House Lane). Routes 48 and 49 extended from Muirhead Avenue Bridge to Muirhead Avenue East with peak hour extras to Lower Lane and Gillmoss. "Rode along new extension opened today 295/988." (EAG) A special minimum fare 29 left Old Haymarket at 7am and eventually went all the way to Kirkby. Other minimum fare cars were operated on some routes to deter 'hoppers' taking short rides.

The layout at Lower Lane was upgraded to feature a roundabout circled by tracks, as seen here in October 1946. This was a perfect example of modern light rail construction with provision on the roundabout for cars to layover without disrupting the through services. Note the steam waggon in the right hand corner. *[Below] A K Terry/National Tramway Museum*

26 October 1943: Routes 13, 19, 29, 44 and 48 extended beyond the City Boundary on rural roadside reservation to Hornhouse Lane. Concrete sleepers, ballast made largely from wartime rubble and more second-hand poles from Coventry were used.

This scene captures the rural nature of the line with the tracks running past wheat fields and farms with names such as Shrog's, Old Ship and Cherry Wood. *[Above] Phil Tatt/Online Transport Archive*

23 December 1943: Extension to Kirkby Main Gate (also known as 'Admin') is opened. This included a curving roadside reservation flanked by more fields and farms. Off-peak, cars only ran every half hour but as many as 3000 an hour were transported at shift-change times. "Rode 942 to Kirkby on 48 on new extension opened today – open country – cows – fields – vegetables. Fast run, non-stop – not a soul in sight – crowds board for the trip back." (EAG)

1943: Experimental car 44 scrapped during the year.

9 March 1944: 'Admin' loop opened. Located outside the ROF Main Gate at Kirkby, this double-track terminus had a Bundy Recording Clock, shelters and assigned queuing points for routes 13, 19, 29, 44 and 48.

12 April 1944: Liverpool's final extension served ROF No 5 Gate at Kirkby. Only used at shift change times, the half-mile roadside reservation had no intermediate stops. The stub terminus had some basic shelters and another Bundy clock. At key times, traffic warranted the use of inspectors at both Admin loop and No 5 Gate (usually known simply as '5 Gate') to regulate departures.

May 1944: To eliminate an awkward length of single track, a short section at the southern end of St Oswald's Street was doubled. *[Right] F N T Lloyd-Jones/Online Transport Archive*

8 September 1944: 943 overturned on Ainsworth Lane curve on the East Lancashire Road. Once righted, the remains were towed away by another tram.

3 November 1944: 814 heavily damaged after colliding with an army lorry at the junction of James Street and the Dock Road.

28 November 1944: Knowing that the end of the war was approaching, the Transport Committee asked Marks to present a preliminary report on the Post-War Reconstruction of the undertaking. One of Mark's options was further modernisation using fast, modern single-decker trams similar to those operated in some US cities. Following a lengthy discussion, he was asked to submit a detailed report with costings.

Background and wartime | 17

Tram fleet

On 8 August 1944 there were officially 742 trams in stock with some inactive or in the snowplough fleet.

Fleet numbers	Dates	Notes
502, 506, 507, 513, 527, 535, 539, 540, 542, 543, 544, 546, 550, 553, 555, 558, 561, 564, 566	1908-11	Known as 'Bellamys' after the General Manager who designed their upper deck roofs, these 64 seat cars had two 40hp motors. Only 544 in passenger use. Some served as snowploughs and a few of the former ARP cars may have been dismantled.
571, 577-602	1913-20	Powered by two 40hp motors, these former double-staircase cars had all been rebuilt by 1924 with a single staircase at either end. After the war, some were fitted with vestibules. 588 was odd man out. It was totally enclosed in 1936 and given an EMB flexible axle truck, 60hp motors, air brakes and large indicators. 597 listed as a snowplough.
614, 618, 621, 625, 627-631	1919/20	These were from a group of English Electric built reversed-stair, 64-seat open-fronted, open balcony cars with two 40hp motors.
43, 117, 122, 134, 303, 304, 306, 314, 322, 330, 335, 363, 603-605, 607, 608, 634-636	1921/22	A motley collection of unvestibuled cars with canopied balconies although those on 603 and 604 were enclosed in 1925. In 1942, 608 was totally-enclosed and 122 and 303 became snowploughs. 322, 330, 634 and 636 were eventually fitted with vestibules.
1-149, 301-471, 572, 637-720	1922-33	With gaps in the numbering, most were conventional four-wheelers designed and constructed by the Corporation when Percy Priestly was General Manager (1920-33) and as such were often referred to as 'Priestly Standards'. Although they had a wide range of trucks and equipment, most had seats for 64 and two 40hp motors. 358 of 1922 was the first to have an enclosed top deck. Subsequently, many were built as totally enclosed.
637-720	1924-26	When new, this batch mostly rode on long-wheelbase radial trucks which it was hoped would improve riding quality. All were reconditioned in the 1930s most being fitted with vestibules and improved seating.
721-732	1927/28	These 'Lengthened Standards' had 20ft saloons and rode on Corporation-built radial trucks of differing lengths, one being as long as 11ft 6in. Also, when new, 726 had seats for 78. In the late 1930s, all were modernised with EMB flexible-axle trucks, air brakes, 60hp motors, improved seating, vestibules and, in some case, larger indicator displays
733-744	1927/28	This small batch reverted to the earlier, standard 16ft 6in body with shorter wheel-base trucks, most receiving post-war windscreens. Although Priestly and Marks disapproved of glass windscreens on safety grounds, 742 was the first totally enclosed car in the fleet.
745-756	1927/28	Originally, these 71-seat Lengthened Standards with 20ft bodies had radial trucks and two 40hp motors. However, in the late 1930s, they were totally enclosed and given EMB flexible axle trucks, 60hp motors, air brakes and, in most cases, improved destination displays
1, 6, 56, 60, 71, 93, 111, 112, 123, 146, 149, 307, 309, 313, 315, 319, 331, 356, 372, 383, 387, 389, 400, 410, 429, 432, 441, 642, 681, 701, 744	1935	To improve overall performance, selected Standards were given more powerful motors and magnetic track brakes although the latter were subsequently removed. The new motors tended to place additional strain on the timber-framed bodies.

Fleet numbers	Dates	Notes
440 (and others, see notes)	1935-39	440, which dated from 1932, was thoroughly modernised in 1935 with a Maley & Taunton swing-link truck and airbrakes and, as such, set the pattern for a further 76 'Pullmanised' Standards including: 5, 12, 28, 31, 35, 38, 54, 81, 87, 88, 89, 91, 95, 96, 98, 101, 105, 114, 126, 128, 147, 305, 307, 316, 317, 318, 328, 329, 334, 336, 338, 340, 342, 343, 353, 367, 382, 385, 386, 391, 393, 407, 420, 445, 451, 454, 459, 469, 588, 673, 721-732 and 745-756. All had EMB flexible axle trucks, 60hp motors, improved lighting and seating, illuminated trafficator arrows and a power brake warning triangle on the dash. Most also had enlarged destination displays.
758-769	1931/32	Built at Edge Lane, these 70 seaters had 21ft bodies and English Electric monomotor bogies. However, these proved problematic so between 1938 and 1944, nine were given EMB lightweight bogies and four 40hp motors. Of these 760, 762, 764. 765 and 766 had English Electric remote control equipment and 759, 761, 767 and 769 direct control although their powerful motors proved too much for their controllers leading to overheating. 758, 763 and 768 were withdrawn in 1936 and never upgraded.
770-781	1933	Painted in a new green and cream livery, these 70-seat cars were swiftly nicknamed 'Green Goddesses' a name subsequently applied to all new trams. They had EMB HW/1 trucks, a development of a type under newer London trams, and four 34hp motors (some reports give four 27hp). 770-777 had flat as opposed to the elliptical roofs on 778-781.
782-817	1933-35	These 70-seat bogie cars were designed by the City's Electrical Engineer, P J Robinson who adopted the body design of London County Council Tramways No 1 and negotiated a contract with EMB to supply trucks. The cars had platform doors, fold down steps, reversed stairs, upholstered seats, concealed lighting, electric bells and a separate driver's cab which gave them the nicknames 'Cabins', 'Robinson Cabin cars' or 'Cabin Cruisers'. They were 36ft long, weighed 18½ tons and, with the exception of 809 which had EMB LW/1 bogies, they rode on London-style EMB HW/1 (Jo'burg) trucks. Although powered by four 34hp motors (some reports say four 27hp), they tended to be sluggish with poor acceleration. 786 was destroyed during The Blitz but was not rebuilt whilst 795, which burnt out on Mount Pleasant, emerged from Edge Lane as a 'Marks Bogie'.
795, 818-867	1935/36	On taking up his appointment, Marks simplified the Cabin car design. Known as 'Marks Bogies' or 'Pneumonia cars' owing to the draughts whistling round the platforms, this group had ordinary stairs, sliding bulkhead doors, no driver's cabin or platform doors, different destination displays and four 34hp (some sources say four 27hp) motors. Nos 818, 820-842, 854 and 866 had EMB HW/1 (Jo'burg) trucks and 843-865 and 867 EMB LW/1 trucks. Nos 838-867 also had different roof profiles
868-992, 151-188	1936/37	These handsome 78-seat streamlined bogie cars ('Liners') were designed R J Heathman. All but one had four 40hp motors and were mounted on three different types of truck. 868-878, 880, 953, 954, 956, 958-980, 983-992 were on EMB HW/2 trucks, 879, 881-917, 943-952, 955, 957, 981, 982, 151-188 on EMB LW/2 trucks and 918-942 on Maley & Taunton swing-link trucks. 905 was unusual in only having four 34hp motors. 171 and 989 were lost in the 1942 fire at Green Lane depot.
201-300	1938-42	To reduce overall costs, 100 cheaper four wheel versions of the Liners were built. These 70-seaters had two 60hp motors and EMB 9ft flexible axle trucks. They also incorporated some equipment, such as controllers, recycled from older cars. Known as 'Baby Grands', all were not in service at the same time as 299 and 300 entered service after 225 and 228 had been destroyed.

Liverpool did not have a large number of dedicated works vehicles and those that did exist by August 1944 were nearly all former passenger cars. Furthermore, from the outset, the City Engineer and Surveyor's Department was responsible for all matters relating to the track. During wintery conditions, road vehicles were used to clear snow and ice although the Transport Department was responsible for keeping the reserved tracks clear. This list does not include trams assigned to snow clearing duties following the bitter winter of 1941 for which the Corporation proved ill-prepared. Nor does it include cars assigned to ARP and other war-time roles. These are detailed in the main part of the story, along with the subsequent history of many of the cars listed here.

Stores vans

Mostly used to transport items including dried sand between the works and the various depots, all of which were later equipped with their own sand-driers.

S1	Formerly L[rd]1 and possibly EL1. Dating from 1900, this purpose-built, two-axle car with sliding doors on both sides was built at the Corporation's first works on Lambeth Road. It was scrapped in 1947.
S2	Formerly L[rd]2. This former 'Ringbahn' motor car 418 of 1898 was converted into a Cash Van in 1905. Together with S3, it delivered tickets and collected cash from various cash offices located around the system. From 1907, the money was taken to Hatton Garden where the cars used two sidings located beneath the building. When this service was motorised in 1921, S2 became a general stores van.
S3	Formerly L[rd]3. This was another German motor car, 424, converted into a Cash Van in 1905 (see above).

Note: L[rd] = Lambeth Road; EL = Edge Lane.

Tool van

429	This German trailer was located at Garston depot.

Railgrinders and scrubbing cars

Liverpool track was plagued by corrugation and needed regular grinding.

PW4	Formerly S4. This ex-German motor car, 416, was converted into a van to tow railgrinding trailers. In the mid-30s it was given the lower deck from Bellamy 569
PW5	In 1929, former Bellamy 429 was cut down into a single-deck scrubber car.
PW6	In 1935, 332 of 1901 became a single-deck works car. It had a Brill 21E truck and two 40hp motors.
PW7	In 1935, 422 of 1901 became a single-deck works car. It had a Brill 21E truck and two 40hp motors.
PW8	In 1935, 'Little Emma' 126 of 1900 which had been extensively rebuilt in 1921 was cut down into a works car. It had a Brill 21E truck and two 40hp motors.
PW9	In 1935, 'Little Emma' 117 of 1900 was also rebuilt in 1921 before being cut down into a works car. In 1939 it was given the lower saloon from Bellamy 538.

Note: PW = Permanent Way

Overhead Lines Department

Various lorries and petrol-engined tower wagons were used including some road-rail examples which could also be used on the 'grass tracks'. These proved useful during construction of miles of reserved track extensions in the late 1930s and early 1940s and remained active until replaced after the war by ordinary pneumatic tyred tower wagons which regularly accessed the private reservations.

CHAPTER 2 | 1945

After five years of neglect, trams and track need urgent attention. Too many newer cars laid up leaving older, more robust veterans to soldier on.

With increasing prospects for peace, the City Council looked towards rebuilding the damaged city. This view shows the devastation on the south side of Lord Street (left), one of the city's principal thoroughfares. In an interview for the *Liverpool Echo* on 17 April 1944, Marks had stated "the future of the tramways is closely linked up with ambitious new road projects. We certainly aim at getting express services to the suburbs. When the new roads come it will be a comparatively easy matter to lay down specially-prepared tram tracks for fast traffic purposes." He had also been considering the next generation of new trams and drawings exist of a four wheel version of the Marks Bogie as well as a centre-entrance double-decker with state of the art trucks and equipment. "The tram of the future will be a streamlined affair, much quieter and faster, and designed to give more comfort for the passenger." Aware that the City Engineer tended to overlook transport concerns, Marks did warn such plans were in the future and not entirely within the scope of the Transport Department. *[Below] Liverpool City Engineers and Surveyors*

On 16 October 1945, Marks submits a costed report to the Transport Committee.
1. Retention and modernisation of the tram system with subways and modern single deck cars (£7,439,360)
2. Retention and modernisation but with current streamline type cars (£6,765,400)
3. Replacement by trolleybus (£4,424,700)
4. Replacement by bus (£3,779,450)

Some aspects were far-sighted looking 20 years into the future and included costs for planned reserved track extensions and for relocating 38 miles of street track onto reservation. However, the report was somewhat tentative on the long-term future of light rail. This may have been because Marks knew which way the wind was blowing. After considering the options, both political parties voted for the cheapest option, their decision being ratified by the full Council a month later, again with virtually no opposition. No consideration was given to the concept of operating mixed modes.

Marks estimated the conversion programme would cost £4,000,000 and should take place in three stages. Stage One (1948-52) would eliminate 28.93 route miles including worn track, single track and loop sections, awkward layouts as well as all mileage within the Borough of Bootle and Litherland UDC. It would also release life-expired vehicles for scrap. Stage Two would occur during 1952-55 and Stage Three during 1955-58.

With the zeal of a convert, Cllr G W G Armour, Chairman of the Transport Committee, who had previously supported tramway retention, now declared no one wanted a city with wires and retaining the trams would hinder construction of new roads so the 'flexible' bus was the way forward. Although the local press were generally supportive, the *Daily Post* did raise concerns about higher fares whilst also stating "the long bus v. tram controversy may soon be ended" and that "the changeover from trams to buses is, of course, in accordance with expert views all over the country."

The knowledge that a future Labour Government would nationalise the electricity industry may also have influenced the Council as the trams relied on relatively inexpensive electricity generated by the Corporation-owned coal-fired power stations at Clarence Dock and Lister Drive.

The wholesale conversion was opposed by the newly formed Liverpool Tramway Passengers' Association under its President Miss Barbara Hoyle. Affiliated to the Light Railway Transport League, which campaigned for modern light rail, the LTPA tried every means to raise legitimate concerns with rate-payers, councillors, press, trade unions, licensing authorities and the Liverpool Corporation Passenger Transport Department. However, out of a population of nearly 750,000 only 1500 signed a petition opposing the policy.

To handle the volume of traffic at Pagemoss, the old crossover was replaced by sophisticated new layout including through tracks plus a turning circle and a double track siding (seen here) with two crossovers on the north side. A planned cross-country extension to Dwerryhouse Lane was never built. Although delighted with the new arrangement, local expert Jack Gahan (JWG) noted ominously that "the track beyond to Prescot in appalling condition and needs immediate attention."

[Above, right] Photographer unknown/National Tramway Museum

"After the war, most trams were still allocated to the same depots as in 1939. The 163 Liners were mostly at the three south end depots with 36 at Green Lane, 4 at Walton and just one at Edge Lane. None appear to have ever been assigned to Litherland." **Stan Watkins**

22 January 1945: Out of a platform staff of 3800, 740 failed to turn up leaving 70 trams idle in their depots. Staff absenteeism and shortages continued to be major problems, especially when the women conductors were all replaced later in the year by men returning from various wartime roles.

24 January 1945: Whilst working an inbound 10B on an icy night, the brake valves on Liner 910 froze at approximately 7.30pm, resulting in a sudden loss of air pressure. The car careered down London Road gathering speed on the steep section of William Brown Street before hitting Marks Bogie 850 side on as it turned from Byrom Street into Dale Street on route 27. The impact sent the latter crashing against a traction pole and a shop. One man was killed and over 70 injured some of whom were carried to nearby empty trams. A mile-long queue of trams built up along Scotland Road but diversions were quickly implemented and normal service was resumed by midnight. At the subsequent enquiry, the Ministry of Transport inspector blamed 'extreme weather'. Buzzers were fitted into the Liners to warn drivers of air pressure loss, and notices issued to drivers.

22 | THE LEAVING OF LIVERPOOL

Automatic Electric Braking (AEB) kits were purchased but, for some reason, not all were fitted. *[Opposite, bottom right and above] Leo Quinn collection/Online Transport Archive*

April 1945: Precast concrete shelters approved for use on more exposed reserved tracks offered little protection. To reduce structural damage in collisions, a new heavier type of bumper was fitted to the Liners starting with 892, 897, 157 and 916.

12 April 1945: Fire at Walton depot. Baby Grand 217 destroyed and eight cars scorched. Roof, wooden cleaning galleries and overhead wiring troughs badly damaged. Known to staff as 'Spellow' and shown on screens as 'Spellow Lane' or 'Walton,/Spellow Lane', this large complex was in two sections.

14 April 1945: Fog was a recurring hazard. Whilst working an early Saturday morning duty, 913 was badly damaged when it collided with a fully-laden contractor's coach at Kirkby. Three people were killed and 12 seriously injured. The mangled remains were towed away by another tram. *[Above] Photographer unknown*

8 May 1945: VE Day marked the end of the War in Europe but all too quickly the sterling work undertaken by the trams and their crews was forgotten. On top of scheduled duties, they had transported and, in some instances, distributed food to over 4,500,000 Allied troops, and repatriated Prisoners of War, evacuees and internees. Sometimes up to 200 trams could be involved in night movements on behalf of the military.

"I remember the VE day parade – standing on an air-raid shelter (I think) waving my Union Jack but wanting it to be over so I could go back to riding and watching the trams. I lived in Wallasey and most Saturdays my mother and I would take the bus to Seacombe where we caught the ferry to the Pier Head. On arrival my mother would go shopping. From about the age of seven, I was allowed to go tram riding initially from South Castle Street. My favourite routes were the 26 and 27 – an hour round. When I got back my Mum would be waiting. Then I branched out and tried to ride all the routes from 1 upwards. I wish I could remember more about these journeys. I do recall being frustrated by not finding a number 7. One ride sticks in the memory. I went to Seaforth on the LOR. Went down to the tram terminus and boarded an 18 which took me away from the Pier Head dipping and ducking over the worn tracks in Bootle. I know now I ended up at Breck Road. Managed to work out I needed to catch a 13 or 14 back into town. I had ridden both these routes a few months earlier. Not sure whether I actually reached 49!" **(MJ)**

10 June 1945: 588 is badly damaged in a collision with 898 on Prescot Road.

3 July 1945: 281 burnt out on Robson Street. Here it is seen at Seaforth on 26 March 1944 *[Below] A G Wells*

1 August 1945: Death of J S Ross, Rolling Stock and Works Engineer since 1934, who was active in the design and construction of the modern tram fleet.

10 August 1945: Another Baby Grand 209 goes up in smoke, this time at Woolton. None of these badly damaged modern trams were officially written off until 12 November 1946.

6 November 1945: 259 damaged beyond economic repair following a collision with a US Army lorry on Edge Lane.

19 December 1945: 809 wrecked in a head-on collision with a US Army tank recovery vehicle on County Road. Sometimes such accidents were caused when drivers of US military vehicles were on the wrong side of the road.

Only a handful of views show the city streets filled with trams. Above, a variety of cars circumnavigate William Brown Street and Dale Street shortly after the end of the war. On the right is a line up of four cars on Lime Street, which had some ten regular services but was also served at peak hours by many extras. Liverpool was still very much a tram city. *[Above] Martin Jenkins collection/ Online Transport Archive; [Right] A M V Mace/National Tramway Museum*

"It is hard now to imagine the thousands of people carried each day on the trams. I remember very lengthy queues were common, particularly during the evening peak hour. Only a few posh people had cars so nearly everybody else had to rely on public transport just as all sorts of shortages began to bite. It was clearly difficult to simply keep everything on the move." **(MJ)**

24 | THE LEAVING OF LIVERPOOL

CHAPTER 3 | 1946

In November 1945 *Modern Tramway* printed an aggrieved editorial 'Whither Liverpool?' condemning the decision to scrap the trams as "a false economy, ill-conceived, ill-considered, and if acted upon utterly disastrous". The magazine urged the Council to reconsider and for concerned ratepayers to protest. Alderman Shennan of the Conservative Party appeared to back-track stating on 24 November 1945 that "all trams may not be scrapped and that in my view the routes to Kirkby, Knotty Ash, Fazakerley and Roby will not be scrapped for some time – if at all."

Unfortunately, the average citizen showed little interest. Their city was battered and dirty. Houses, food, coal, clothes and shoes were in short supply. For many, the war-weary trams seemed synonymous with post-war austerity. The future lay with new buildings, slum clearance, planned housing estates, suburban shopping malls and 'flexible' buses.

As a result of the bombing and relocation:
Civilian population of City of Liverpool dropped from some 856,000 in 1939 to around 666,000 in 1944.
Civilian population of the Borough of Bootle dropped from 76,000 in 1931 to some 60,000 in 1945.
Hundreds of prefabricated houses (pre-fabs) quickly erected to replace war-damaged or destroyed property.
Thousands were reliant on over-stretched public transport although the bombing, especially of the northern suburbs and in Bootle and Litherland, led to a drop in traffic on some routes.

Excluding depots, there were 191 miles 483 yards of track including:
8 miles 364 yards in the Borough of Bootle
1 mile 380 yards in Litherland UDC
4 miles 1455 yds in Huyton-with-Roby UDC
5 miles 117 yards in Whiston Rural District Council
1 mile 1340 yards in Knowsley Township
1595 yards in Prescot Township

Route mileage totalled 97 miles 946 yards of which:
27 miles 1508 yards were on reserved or 'grass' tracks and 6 miles 1389 yards were in and around depots, with Edge Lane having the lion's share at 2 miles 10 yards.

Approximately 740 cars in stock housed at seven depots with combined undercover space for 662. However, many cars required urgent attention but Edge Lane Works seemed unable to cope; for example 884 was at the back of Garston depot for five years together with 253 which had 'dropping ends'. Some problems highlighted in Marks' 1936 report remained unresolved. To make matters worse, there was a shortage of skilled men, vital spare parts such as motors, armatures and field coils and an inbred reluctance to engage outside contractors. Also more time and energy was being given to the growing bus fleet.

This view of a group of Control Officers was taken outside the Transport Department's Headquarters at 24 Hatton Garden, known to employees as 'Twenty-Four'. The building survives today as the Richmond Hotel.
G W Rose/Online Transport Archive

Some 8000 people were employed by the Department. Many men aged between 65-70 were still working as inspectors, cleaners, fitters and mechanics, and 'floating' or 'pavement' conductors who sold tickets at queues or helped on overcrowded vehicles. *G W Rose/Online Transport Archive*

On the positive side, the CE&S was renewing worn sections of rail. Between 1944 and 1946, Belvedere and Devonshire Roads were relaid and work began on renewing significant stretches of reserved track on Mather Avenue. Concrete sleepers were used and the ballast was a concoction of broken brick and red shale. The latter gave a more solid foundation than the former use of earth and ash which lacked sufficient drainage outlets often reducing the grass tracks to quagmires during wet weather, which also caused rail joints to drop. During the relaying, trams kept running using single track sections with temporary crossovers controlled by electric signals. *[Above] F N T Lloyd-Jones/Online Transport Archive*

Other urgent relaying and patching took place at the Pier Head, Water Street, London Road, Green Lane and St Anne Street. This view shows similar labour-intensive work being undertaken at Pier Head in the early 1950s. *[Below] G W Rose/Online Transport Archive*

January 1946: 507 and 566 joined the snowplough fleet.

February 1946: First cars painted in new green and white livery were 607, 614, 78 and 660. English Electric balcony car 614 of 1919 is seen in these new colours at Old Haymarket. Although decidedly old-fashioned, the Preston-built bodies were still structurally sound. *[Above] N N Forbes/National Tramway Museum*

Bowing to further Union pressure, a start was made on fitting vestibules to the remaining open-fronted cars. Some with old style handbrake columns projecting beyond the dash had these replaced. Marks now opposed the vestibules on cost grounds. Many vehicles would soon be scrapped and as far as the driver's complaints were concerned – 'it wasn't cold and didn't rain every day!' Despite protective clothing, life 'up front' was tough for drivers on unvestibuled cars who often faced the full force of the weather. One former driver said "You needed the constitution of a polar bear." Another, Tom Webster, who drove the city's last tram and joined the Department in 1919, recalled: "I used to keep warm by getting my wife to sew brown paper inside my

26 | THE LEAVING OF LIVERPOOL

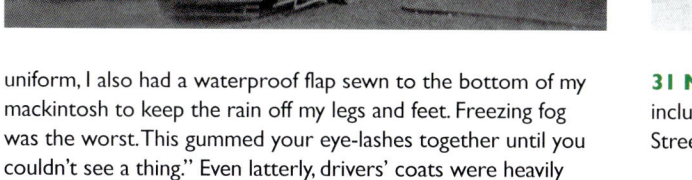

uniform, I also had a waterproof flap sewn to the bottom of my mackintosh to keep the rain off my legs and feet. Freezing fog was the worst. This gummed your eye-lashes together until you couldn't see a thing." Even latterly, drivers' coats were heavily quilted. *[Above, left] J W Gahan*

The majority were fitted with a crude affair of glass and plywood. Nicknamed 'Prefabs', they were attached to the canopy with the gap between the dash being covered by a strip of waterproof canvas. The work was carried out at the Works, as well as at Dingle and Prince Alfred Road depots and was completed by 1948. *[Above, right] J S Webb/National Tramway Museum*

11 March 1946: To alleviate congestion caused by peak hour trams reversing in the middle of Byrom Street (seen here), a queuing point was established on Great Crosshall Street. Used by routes 2, 22, 43 and 44 it was quite convenient for the central business district. *[Below] Martin Jenkins collection/Online Transport Archive*

31 March 1946: By now, 12 major junctions had been relaid including Everton Valley/Netherfield Road, Lord Street/Church Street/Whitechapel, St George's Crescent/Preeson's Row.

4 April 1946: The trams really came into their own on major sporting occasions when they moved thousands swiftly and efficiently. This was especially true during Grand National Week when hundreds of extras were provided usually showing 'Races' on their screens. On Grand National Saturday, Islington was closed to trams and used by special buses heading for Aintree. Extra trams on routes 20 and 21 were supplemented by a shuttle service from Walton, some of which terminated at Hall Lane or Warbreck Moor. "Aintree bound buses crawled along mixed up with a continuous stream of motor cars, the trams just sailed past, the tram tracks being kept clear." (EAG) On arrival, some returned to Walton depot whilst others were lined up side-by-side and bumper-to-bumper on both tracks as seen in this view of 880. When ready to depart, the cars working 'wrong line' used the crossover at Warbreck Moor which would also be used by service cars on the 20 and 21. *[Above] E A Gahan/Online Transport Archive*

April 1946: The first of six Guy Vixen tower wagons was delivered to add to a similar vehicle that had arrived during the war. The bodywork was built at Edge Lane and consisted of a small workshop and mess room area, behind which was fitted a three-stage, hydraulically-operated tower with a revolving platform. The body contained parts salvaged from trams such as seats for the crew and a gong for signaling to and from the tower. They were amongst the first vehicles in the fleet fitted with mobile radio. All overhead had to be regularly inspected, probably once a week. *[Below] G W Rose/Online Transport Archive*

20 May 1946: Bowing to political pressure, Marks reluctantly agreed to reintroduce routes 30 and 31 which he believed would be unprofitable as they now served neighbourhoods depopulated by the bombing. Operating every 30 minutes, both connected the Pier Head to Walton (Spellow Lane). Leaving from the North Loop, the 30 ran via Dale Street, Islington, Shaw Street and Netherfield Road. The 31 travelled by way of Lord Street, Church Street, Lime Street, London Road, Brunswick Road/Erskine Street, Everton Road, Heyworth Street and St Domingo Road. In peak hours, some journeys extended north to Walton Park and Walton Church. Here, 227 tackles St Domingo Road, one of the steepest grades on the system. Note the centre poles with their warning black and white hoops and the fascinating array of advertising hoardings concealing bombed sites.

[Above] N N Forbes/National Tramway Museum

May 1946: A few streamliners were painted in the pre-war livery complete with white roof and curved fleet title circling the coat of arms. However, most emerged in the new green and white livery.

28 | THE LEAVING OF LIVERPOOL

May 1946: To alleviate overcrowding on the Pier Head's South Loop, a new siding was laid parallel to the main approach track from Mann Island and was used as boarding point for the 8s, 33s and 45s. *[Above] Martin Jenkins collection/Online Transport Archive*

10 July 1946: 5.20pm. Trolley dewires on outbound car 960 at Knotty Ash. Standard 79 pulls up behind but is hit by 961 which catapults 79 into 960 causing extensive damage. All three packed with home-going city-workers, many of them strap-hangers, and 28 were injured. There was confusion for several minutes but drivers and conductors, though themselves injured, assisted passengers from the wrecked vehicles. *[Right] Stan Watkins collection*

22 July 1946: Routes 23, 24 and 28 terminate at Old Haymarket as opposed to Lime Street Quadrant on weekday evenings, as well as all day Sunday and after 2pm on Saturdays. On the left, defective car 28 is being examined. Note the cardboard '30' in the driver's windscreen. *[Below] Roy Brook/National Tramway Museum*

17 August 1946: "9.30am. Car 712 on route 12 runs away down Erskine Street and rounds corner into Moss Street after having hit rear of 685 (route 46). 712's brake gear snapped. 712's front end came off. 685 also much damaged." (JWG) Both cars were repaired.

Autumn 1946: Townsend Avenue relaid with concrete sleepers and red shale for ballast.

19 August 1946: A new range of simplified Bell Punch tickets was issued. *[Right]*

30 September 1946: Nearly 200 cars off the road leaving 545 to meet the scheduled peak turnout of 601. 83 were laid up at Edge Lane with a further 109 out of service in various depots. As a result, there were major service reductions, fewer peak hour extras and long queues of disgruntled passengers. Looking down-at-heel and still covered in its drab wartime brown wash, 871 speeds towards Garston along the Horrocks Avenue reservation opened in 1939. *[Below] E N C Haywood*

To counter public dissatisfaction with the level of service on some of the busier corridors, parallel bus routes were introduced to provide additional capacity although, in a sign of things to come, workmen's return tickets were not available. A policeman eyes the photographer as he takes this view of 607 on an Old Haymarket evening extra to Lower Lane. In the background is 896 which will later be consumed by fire at Green Lane depot. *[Opposite, top] N N Forbes/National Tramway Museum*

October 1946: When the loss-making, peak-hour service 11A (Pagemoss to Pier Head via Green Lane) was discontinued, a new unnumbered service was introduced between Pagemoss (later Longview Lane) and Caird Street which catered for those employed at the giant Ogden Imperial Tobacco factory on West Derby Road. On 19 February 1947, 713 is seen on Green Lane. Although signed for 'Sheil Road', it was probably en route to Caird Street. Shortly, after being fitted with a vestibule in June 1947, 713 was another burnt in a fire at Green Lane depot. Extras on route 29 also linked Caird Street to the housing estates flanking Muirhead Avenue. *[Opposite, bottom left] J W Gahan*

The unprofitable, part-time route 43A (Pier Head, Clayton Square or Old Haymarket to Utting Avenue via Robson Street) was also discontinued. This wartime view of 581 was taken at Pier Head before the ex-Double Staircase car was fitted with a post-war vestibule. *[Opposite, bottom right] N N Forbes/National Tramway Museum*

Inbound cars on the 22 terminating at Great Crosshall Street now showed the number 2 as did outbound cars from Pier Head or Great Crosshall Street terminating at Hall Lane.

1 November 1946: The use of red-on-white route numbers was officially discontinued. The relevant leaflet explained – "in future route numbers displayed will indicate destination, without bearing on its starting point." Although mostly used by cars terminating at South Castle Street, the red-on-white combination was sometimes shown if a car was terminating somewhere short of the Pier Head. The following are known to have appeared in

30 | THE LEAVING OF LIVERPOOL

red-on-white: 1, 1A, 6, 6A, 8, 8A, 9, 10, 10A, 10B, 10C, 11, 11A, 12, 13, 14, 15, 26, 27, 33 and 40 and the colours may also have been used for 7, 9A, 15A, 26A, 27A and 32 but no photographic evidence exists. This view of 327 of 1931 is taken on Lime Street as it heads towards the peak hour queue point at Roe Street (shown as 'Lime St' on screens). *[Overleaf] A W V Mace/National Tramway Museum*

12 November 1946: The sale of the remains of 809, 850, 854, 910, 913 and 943 approved by the Transport Committee. These were duly scrapped shortly afterwards. Looking suitably skeletal, 910 is seen before its final cremation.

[Right] Martin Jenkins collection/Online Transport Archive

November 1946: Following fire damage, 733 was the first Priestly Standard to be withdrawn after the war.

4 December 1946: As the extension along the East Lancashire Road had been granted under emergency wartime powers, the Corporation had to apply for the necessary powers to construct an already existing line! Since the end of hostilities and the sale of the ROF to the Corporation for conversion into a trading estate, the Kirkby route had morphed into something of a white elephant with virtually no off-peak traffic. However, at peak times, cars remained well-patronised even though less than 1000 were now employed at Kirkby.

December 1946: Peak hour journeys from Great Crosshall Street on route 43 now show 43B. More track renewal on parts of West Derby Road, Prescot Road and Everton Road and new junctions installed at Erskine Street/Moss Street and Castle Street.

Following cars overhauled and repainted: 50, 72, 341, 693, 787, 807, 906, 958, 959, 960 and 991.

The following passenger cars believed to have been withdrawn by the end of the year, some of which had been out of service for some time: 589, 590, 595, 604, 628, 631, 634 and 733. It is not known when the cars were actually scrapped. Also officially withdrawn were 564 and 575, neither of which had run for many years. The remains of the following cars were either scrapped at the end of the year or during 1947: 809, 850, 854, 910, 913 and 943.

32 | THE LEAVING OF LIVERPOOL

CHAPTER 4 | 1947

Although the tram system was condemned no closures had taken place. There were now officially 723 cars in stock of which a staggering 200 were still unserviceable, equal to 28% of the fleet. It is not clear where priorities lay at the works and what was the policy regarding urgently needed repairs. Despite a few repaints, the fleet looked uncared for leading the public to associate trams with breakdowns, power cuts, queues and long delays. Most working people relied on public transport. There was still a 5½ day week, rationing was in force, commodities were scarce and few had telephones or televisions. At weekends thousands, simply wanting to escape their drab surroundings, boarded Birkenhead and Wallasey Corporation ferries to explore the Wirral or bask on the beaches at New Brighton. Hundreds visited dance halls, cinemas and theatres or attended sporting events such as football, horse racing, speedway, boxing, greyhound racing and cricket. Church-going still played a key role in everyday life.

Hard-pressed depot foremen struggled to put sufficient trams on the road. Every 'runner', no matter how old or structurally unsound, was kept going.

The Electricity Act of August 1947 led to the nationalisation of the Corporation's two power stations.

When Norman Forbes replaced Gilbert A Jones as the Liverpool area LRTL representative in late 1946 he, and others, were determined to challenge the conversion programme whenever possible.

This was the list of routes published in the March 1947 time-table:

TRAM ROUTES

1	PENNY LANE & CITY via Garston & Aigburth	22	FAZAKERLEY & CITY via Walton Road
1A, 1B	GARSTON & CITY via Aigburth	*22A	FAZAKERLEY & CITY via Hale Road
*2	BLACK BULL & CITY via Walton	23	SEAFORTH & CITY via Strand Road
3	WALTON & DINGLE via Lime Street	24	SEAFORTH & CITY via Knowsley Road
4	PENNY LANE & CITY via Wavertree	25	WALTON & AIGBURTH via Moss Street
4A	CHILDWALL & CITY via Wavertree	26	OUTER CIRCULAR via Oakfield Rd. & Lodge Lane (short journeys 26A)
4W	WOOLTON & CITY via Wavertree	27	OUTER CIRCULAR via Lodge Lane & Oakfield Rd (short journeys 27A)
5	WAVERTREE & CITY via Smithdown Road		
5A	PENNY LANE & CITY via Smithdown Road	28	LITHERLAND & CITY via Scotland Road
5W	WOOLTON & CITY via Smithdown Road	29	EAST LANCASHIRE ROAD & CITY via Tuebrook
6	BROADGREEN & CITY via Edge Lane	29A	MUIRHEAD AVENUE EAST & CITY via Tuebrook
6A	BOWRING PARK & CITY via Edge Lane	30	WALTON & CITY via Netherfield Road
*7	PENNY LANE & CITY via London Road	31	WALTON & CITY via Heyworth Street
8	AIGBURTH & CITY via Garston & Smithdown Road	*32	PENNY LANE & CITY via Park Lane
8A	GARSTON & CITY via Smithdown Road	33	GARSTON & CITY via Princes Park
*9	OLD SWAN & CITY via Kensington	*	LONGVIEW LANE & SEAFORTH via Everton Valley
10	PRESCOT & CITY via Old Swan	*35	FAZAKERLEY & SEAFORTH via Hale Road
*10A	KNOTTY ASH & CITY via Old Swan	*36	LOWER LANE & SEAFORTH via Hale Road
10B	PAGEMOSS AVENUE & CITY via Old Swan	*37	UTTING AV. EAST & SEAFORTH via Everton Valley
10C	LONGVIEW LANE & CITY via Old Swan	*	PENNY LANE & SEAFORTH via Heyworth Street
11	GREEN LANE & CITY via Tuebrook	*39	KNOTTY ASH & CITY via Brookside Avenue
12	WEST DERBY & CITY via Tuebrook	40	PAGEMOSS AVENUE & CITY via Brookside Ave.
13	EAST LANCS. ROAD & CITY via Townsend Lane	*	PAGEMOSS AVENUE & EDGE LANE via Old Swan
13A	BROAD LANE & CITY via Townsend Lane	*	PENNY LANE & EDGE LANE via Mill Lane
14	UTTING AVENUE EAST & CITY via Townsend Lane	43, 43B	UTTING AVENUE & CITY via Everton Valley
*14A	CHERRY LANE & CITY via Townsend Lane	*43A	UTTING AVENUE & CITY via Robson Street
15, 15A	CROXTETH ROAD & CITY via Princes Park	44	KIRKBY or GILLMOSS & CITY via Everton Valley
16	LITHERLAND & CITY via Vauxhall Road	44A	LOWER LANE & CITY via Everton Valley
16A	KIRKDALE STATION & CITY via Vauxhall Road	45	PENNY LANE & CITY via Aigburth & Mill Street
17	SEAFORTH & CITY via Derby Road	45A	GARSTON & CITY via Aigburth & Mill Street
18	SEAFORTH & BRECKFIELD RD. via Everton Valley	46	PENNY LANE & WALTON via Moss Street
*18A	SEAFORTH & EVERTON RD. via Everton Valley	*48	PENNY LANE & GILLMOSS or MUIRHEAD AVENUE EAST & WOOLTON
19	KIRKBY, HORN HOUSE LANE or GILLMOSS & CITY via Robson Street	49	PENNY LANE & MUIRHEAD AVE. EAST via Old Swan
19A	LOWER LANE & CITY via Robson Street	*	DINGLE & GILLMOSS
20	AINTREE & AIGBURTH via Park Road (short journeys 20A)	*	MUIRHEAD AVE. EAST & EDGE LANE
21	AINTREE & AIGBURTH via Mill Street		

*Denotes Part-time Routes

Liverpool's complex transfer system enabled passengers to travel considerable distances on payment of a single fare. This was one of several ideas adopted from the USA and was particularly welcomed by the less well off. Shown right are a few of the official transfer points as detailed in the 1947 time-table.

Two slightly different lists survive detailing the number of cars required for each route – one has a daily turnout of 554, the other 563. The latter also refers to 'The Pool' of 62 spares based at Edge Lane. These were mostly older vehicles listed for disposal but retained for times of peak demand. By early 1947, non-availability was now reaching epic proportions leading to cancellations, bus substitution and overcrowding as seen in this view at Old Haymarket. Some trams are said to have carried over 100 passengers with one visitor claiming to have ridden a tram with "120 crammed on board". *[Above] N N Forbes/National Tramway Museum*

The routes operated by each depot and the turnout on each are shown in the table below.

TRANSFER FACILITIES AVAILABLE ON TRAMCARS ONLY

The Transfer System operates in order to enable passengers to travel from point to point by alternative routes at the same fare as exists on the through routes. Use of the Transfer Facility obviates long waits for through cars in the City area, gives a wider choice of vehicles and avoids the necessity of walking to the terminal points in the City. Generally, the journey must be in the direction of the through route, although exceptions have been made in cases where a through route does not exist, and a link-up permitted with cross-city routes.

THE TRANSFER MUST BE MADE AT A RECOGNISED TRANSFER POINT BY THE FIRST AVAILABLE CAR IN THE REQUIRED DIRECTION.

AIGBURTH VALE
BLACK BULL
BRECK ROAD : Breckfield Road North
BROAD LANE
BROADGREEN STATION
DINGLE : Ullet Road
EAST LANCS. ROAD : Lower Lane
EVERTON VALLEY : Walton Lane
EDGE LANE DRIVE : Broadgreen Road or Mill Lane
GILLMOSS
GREEN LANE : Prescot Road
GREEN LANE : West Derby Road
LINACRE LANE
LONG VIEW LANE
MUIRHEAD AVENUE : Liddell Road
OLD SWAN : Broadgreen Road or Blackhorse Lane
PAGE MOSS AVENUE
PENNY LANE
PILCH LANE
SPELLOW LANE : County Road
STANLEY ROAD : Melrose Road
STANLEY ROAD : Knowsley Road
STRAND ROAD : Stanley Road
STOPGATE LANE
WALTON BRECK ROAD : Robson Street or Donaldson Street
WALTON LANE : Anfield Road
WAVERTREE CLOCK TOWER

Garston		Walton		PAR		Edge Lane		Green Lane		Dingle		Litherland	
Route(s)	Turnout	Route(s)	Turnout	Route(s)	Turnout	Route(s)	Turnout	Route(s)	Turnout	Route(s)	Turnout	Route(s)	Turnout
1/1A/1B 45/45A	23	3	7	4	12	6/6A	29	10	12	15/15A	7	16	4
8, 8A	28	13/13A, 14/14A	38	4A	8	39/40	21	9/9A, 10/10B/10C	56	20/21	21	17	8
33	4	19/19A, 44/44A	43	4W	6	41	*	11	12	25	7	18/18A	19
		22	25	5	11	47	*	12	14	26	16	23/24	10
		25	7	5W	6			29/29A	23	Dingle-Kirkby	*	28	10
		27	18	7	8								
		30, 31	5	32	7								
		35	1	38	2								
		36, 37	3	42	*								
		43, 43B	17	46	8								
				48	*								
				49	7								

* allocation not known

At this time only route 25 was officially operated from two depots (Walton and Dingle) although it seems likely that very early morning and late night duties on the 3 could have been provided by Dingle and those on the 20/21/46 by Walton.

Track wholly or partially relaid during 1947 included lengths of 'grass track' on Muirhead Avenue and Edge Lane Drive together with street track on Allerton Road and West Derby Road. In the city centre Parker Street (down line), Byrom Street, Dale Street and the Ranelagh Street/Lime Street junction were all relaid.

6 January 1947: Following repair of the war-damaged railway bridge at Bankhall, route 16 (Pier Head to Litherland) was reinstated on a 20 minute headway. However, the existing temporary service 16A (Pier Head to Kirkdale Station) continued, also at 20 minutes intervals, giving a 10 minute service over the southern end shared by both routes, although the 16A finished earlier in the evening. There are no known photographs of cars showing 16A. Both routes operated from Litherland depot. Sometime in early 1947, 844 is about to reverse at the temporary 16A terminus outside the Princess Cinema on Selwyn Street. *[Above] N N Forbes/National Tramway Museum*

January 1947: This was one of the worst winters on record. For six weeks, heavy snowfalls required the use of snowploughs and salt cars to keep the tracks clear. The overall sense of misery was heightened by power cuts caused by coal shortages. The extreme cold also led to low voltage which reduced tram speeds and frequencies. "All shopping cars and evening services severely curtailed." (Jack Gahan/JWG) Situation exacerbated by periods of dense smog. Following cars damaged in accidents and placed in store at Prince Alfred Road (PAR) depot, there being no space at the works: 127, 272, 297, 382, 588, 684, 711 and 752. 
[Bottom, left] J W Gahan collection

This is the known snow plough allocation at the start of 1947: 122, 507 (Garston); 502 (PAR); 565 (Dingle); 566 (Walton); 303, 553, 555 (Edge Lane) and 506, 597 (Green Lane). In the first view, 502 is at PAR, still in its pre-war livery of red and cream complete with 'Liverpool Corporation Tramways' on the rocker panel whilst, in the second, 566 is outside the former horse car depot at Walton. Note the canvas screens offering limited protection for the driver, the missing upper deck windows and the absence of indicator boxes on 566. As normal service ended when the last cars left the Pier Head at midnight, it is known that some sympathetic drivers of snowploughs and works cars would give lifts to people where appropriate. *[Below] Martin Jenkins collection/Online Transport Archive; A C Noon/Online Transport Archive*

15 January 1947: Industrial services linking Seaforth to Fazakerley and Utting Avenue East allocated the numbers 35 and 37. Few trams had these numbers added to their indicator screens so cards were sometimes displayed in the vestibule. Some short workings on the 1 were numbered 1B and the peak hour service from Fazakerley to the Pier Head via Hale Road became the 22A. Around this time, another unnumbered industrial service was introduced between Dingle and Stopgate Lane via the 25 and 19 with one morning duty north (6.30am) and one south in the evening (5.05pm).

1947 | 35

16 January 1947: "Travelled on old balcony car 322 (a wreck) from South Castle Street to Green Lane. New bus on route 74 could not overtake on stretch between Lower Breck Road and Tuebrook Station as 322 travelled fast and swayed considerably." (JWG) Here the 'wreck' is seen at Old Haymarket about to depart via St John's Lane on a peak hour working to Broad Green (sometimes shown as Broadgreen or Broad Green or Broad Green Stn on indicator blinds). *[Above] N Forbes/National Tramway Museum*

27 January 1947: "Two cars stationery at Finch Lane (East Prescot Road) run into from rear. Extensive damage to two rear cars. Cars involved 300, 293, 911. Single line working. 911 back in service 28 May 1947." (JWG) This rare view shows 293 after the accident. If only the photographer had recorded 300, the last Baby Grand, of which there is no confirmed photograph. Following this accident, 300 may never have run again as it probably languished at Green Lane waiting to go to the works.
[Right] Leo Quinn collection/Online Transport Archive

9 February 1947: "Balcony snowplough 303 in town, still in red and cream livery with full 'Liverpool Corporation Tramways' on rocker panels. Was in passenger service until 1942. Department so slipshod. Group of men at Pier Head struggling to tie the wooden snow blade on with wire!" (JWG)

March 1947: This quote is from the March-April 1947 Liverpool Transport Guide. "Wot, no trams? Apart from the General Strike, this must have been the only occasion on which so high a proportion of trams mechanically fit for service lay idle in depots all over the system. Plenty of men on hand to work the cars, more people than ever wanting to use them – and less current available to drive them. One irony of the situation is that, whereas during the war a shortage of imported fuel-oil forced us to axe bus services now the war is over it has been lack of home-produced power which produced a tram shortage".

"Twenty minute service on all tram routes after 7pm. 105 trams out of action (mostly streamliners). Old cars working all day and evening services. Six inches of snow resulting in slush flooding motors on the modern cars."
(J W Gahan/JWG)

36 | THE LEAVING OF LIVERPOOL

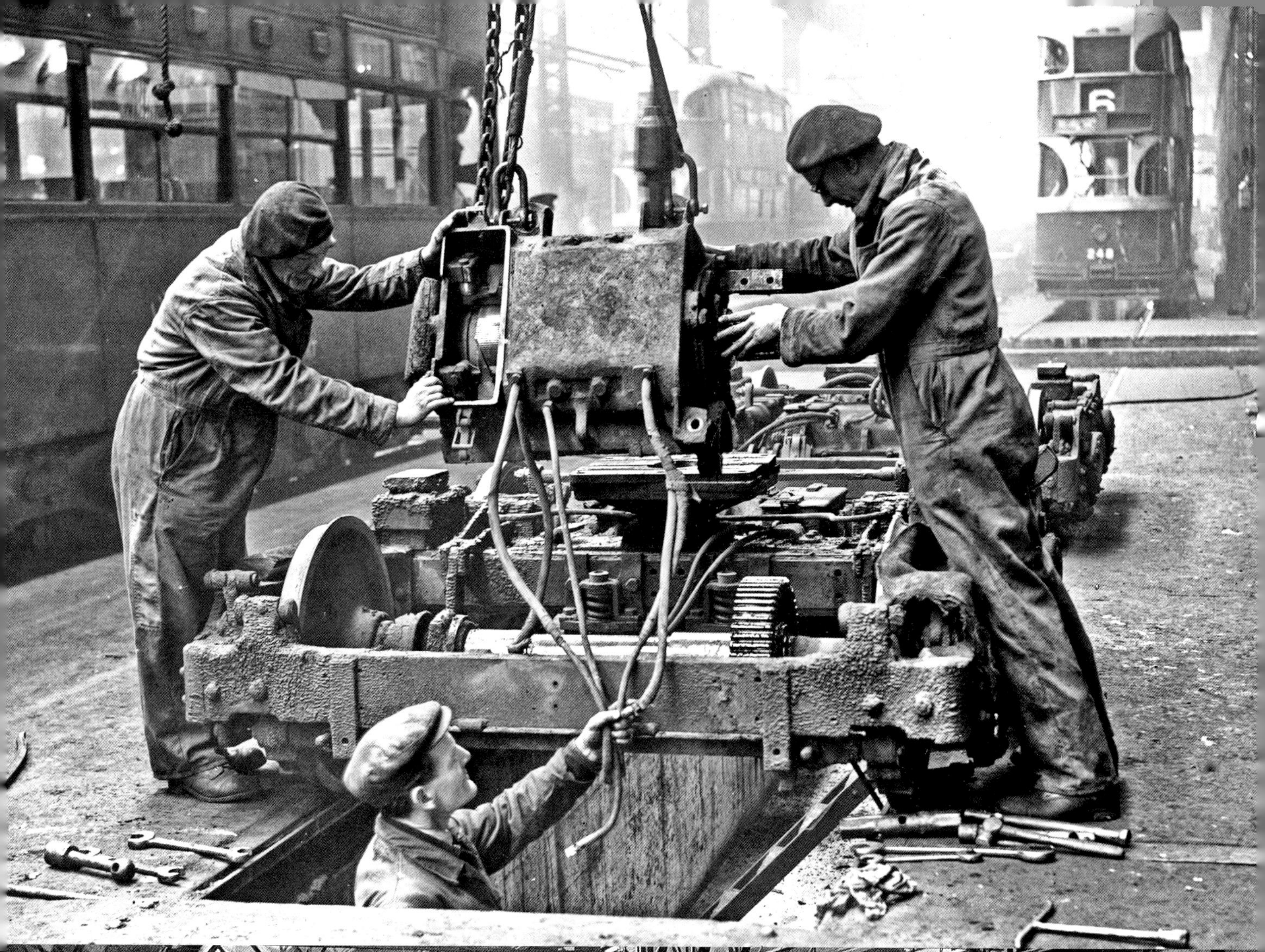
6
248

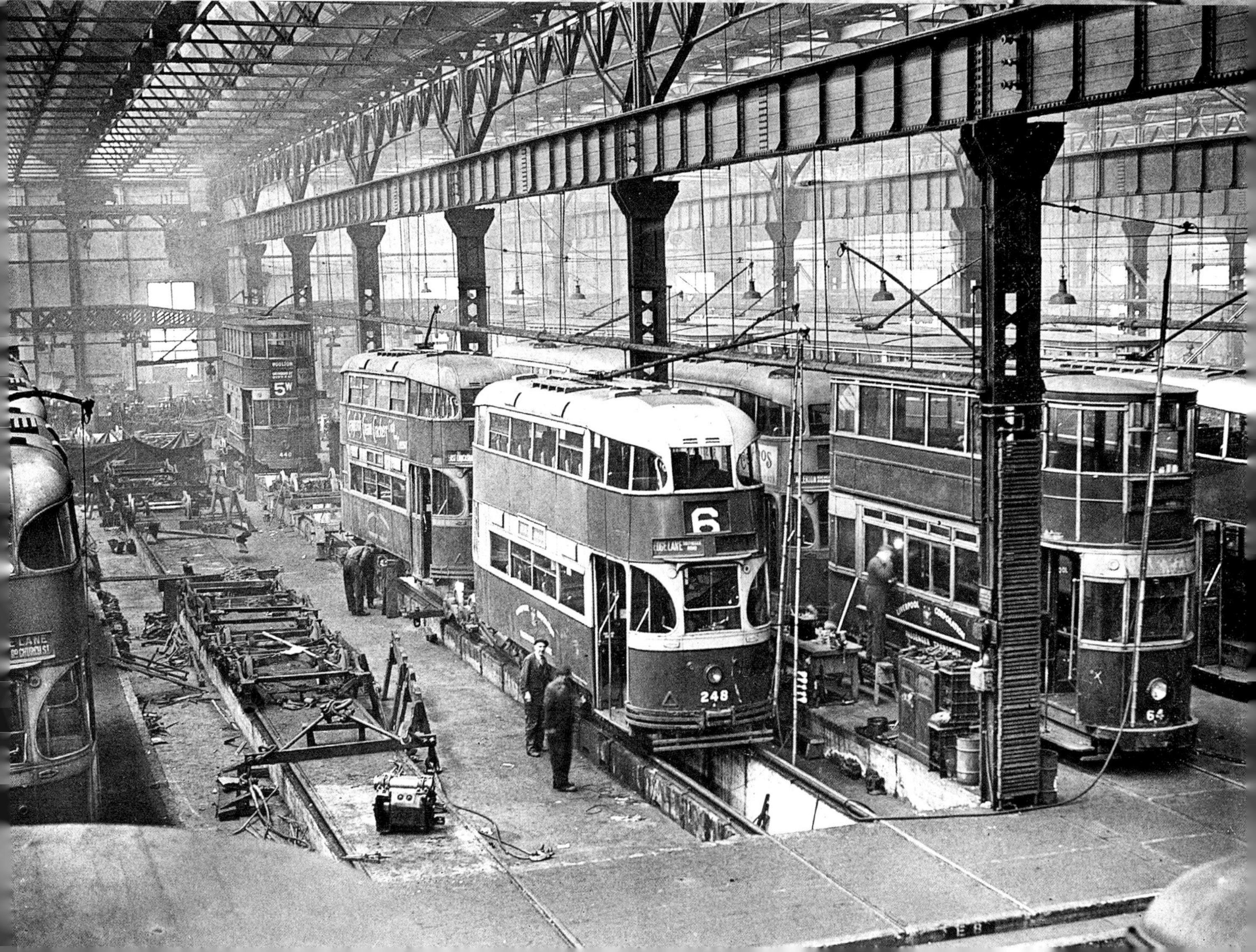
5W
6
248
64

14 March 1947: In the March 1947 time-table introduction, the Transport Department admit that 167 cars are off the road out of a total of some 600. Peak demand was now for 580 but, even at good times, never more than 520 trams available. All this was 'fuel' to the growing bus lobby, which attracted press attention and photos were taken in the works. In the first, a repaired motor is being lowered into position whilst, in the second, cars including 64, 248 and 440 are in the main workshop. At the start of the war, all the windows had been covered and asbestos sheets replaced glass roof panels which were not reinstated until 1950. *[Previous page] Leo Quinn collection/Online Transport Archive (both)*

16 March 1947: So supporters could buy advance tickets for a forthcoming cup tie, special football cars were run from South Castle Street, Earle Road and Norris Green to and from the long siding on Walton Breck Road which stored cars during Liverpool FC home fixtures. At this time, football reflected the sectarian divide in the city. The divisions were so deep some drivers refused to work routes passing particular grounds. "I remember an old tram driver, a real Evertonian. He wouldn't go to Anfield – not even a derby game. When he used to pass Liverpool's ground, he used to spit out. There was another driver at Walton, a big Evertonian. If you were a minute late for work at Walton you lost your car, your car had gone out. You used to hang around the depot in case somebody else lost his. This chap's name was called out and he was told '27' which was always known as the 'belt'. And he said 'I'm not taking that so-and-so car past Anfield. No way.' And he refused to take it. He wouldn't drive a car past Anfield."

22 March 1947: "New posts, crossings and rails for relaying Lime Street – Ranelagh Street junction stacked on bomb site in Great Charlotte Street." (JWG) In an effort to speed up services, some compulsory stops especially at busy junctions were changed to request only.

26 March 1947: After the pile-up on 10 July 1946, 960 was rebuilt and emerged in the pre-war livery. "Out again thoroughly reconditioned, with sliding windows and buffers of new pressed metal design." (JWG) Sliding windows proved more effective than the original wind-down type which tended to jam at jaunty angles letting in rain and creating drafts. Sadly, 960 would go up in smoke less than eight months later. *[Above] C A Mayou*

23-29 March 1947: During Grand National Week extra trams and buses were operated. Starting at 11.00am on 'Jump Sunday' (23 March), special Races cars, charging a special flat fare, departed from Pier Head and Victoria Street. Additional cars were also run on routes 20, 21 and 22. On Race Days (27/28 March), similar arrangements were in force with the addition of a shuttle service between Walton and Aintree. With a full load on board, Walton-based veteran English Electric balcony car 625 passes through Black Bull on its way to Aintree. *[Above] F NT Lloyd-Jones/Online Transport Archive*

29 March 1947: On Grand National Day itself, flat-fare extras left Pier Head, Victoria Street and possibly South Castle Street every few minutes after 10.00am, sometimes in convoy. Additional cars were also provided on routes going to Aintree, Fazakerley and Norris Green as well as on the 13 and 19 to Lower Lane and the 13, 14 and 43 to Norris Green (Broad Lane) where special buses shuttled to and from the racecourse. "Many special trams from Victoria Street to Aintree, including 544. Traffic handled perfectly." (JWG) At the rear of this impressive line-up in Victoria Street is Standard No 8 which was withdrawn in the September just months after being fitted with a driver's windscreen! It was broken up at Dingle the following August. *[Above] J W Gahan*

April 1947: More industrial services assigned numbers. Dockers' specials from Seaforth to Longview Lane and Penny Lane became 34 and 38 respectively whilst the three from Edge Lane, Southbank Road, became 41 (Pagemoss), 42 (Penny

38 | THE LEAVING OF LIVERPOOL

Lane) and 47 (Muirhead Avenue East). Again, few cars had 34 and 38 added to their destination blinds. In this remarkable view taken at Seaforth, Maley 925 from PAR is on route 38, EMB Heavyweight 980 from Green Lane on the 34 and Cabin 791 from Walton the 37. None displays a route number. The 25 Liners on Maley & Taunton trucks only ventured into Bootle on route 38. The spring carrier on their trucks could easily foul on raised setts protruding above the poorly maintained track. The photographer's bicycle stands on the left. *[Above] N N Forbes/ National Tramway Museum*

From some unknown date, cars working inbound from Prescot, Longview Lane and Pagemoss to Old Swan, Green Lane depot or points in the city showed 9A.

18 April 1947: "544 on route 49 in evening peak. This became regular and car did a trip from Penny Lane to Gillmoss and back each evening Mon-Fri." (JWG) Jack's younger brother Ted took this photograph of the car on Muirhead Avenue Bridge making its way back to PAR in the early evening. *[Bottom, left] E A Gahan/Online Transport Archive*

1 May 1947: During the early post-war years, local enthusiast Clifford Noon made several visits to the works. On this occasion, he noted 306, 580, 589, 604, 618, 621, 628, 631 and 733 on the scrap lines at the rear. It is not known when these unwired sidings were laid.

4 May 1947: The first post-war tour organised by the LRTL left St George's Place at 11.30am with Inspector Hesketh from Hatton Garden driving Garston-based 972. During the next few years, Hesketh helped to organise a number of LRTL tours, the Corporation charging very little to hire a car. When a tram was on hire or not in public service it usually showed either 'Private' or 'Special'.

At Garston, participants were able to photograph snowplough 122 and German trailer 429 of 1898, which was towed out specially by Standard 441. A group photograph was also taken.

[Below] L L Quinn collection/Online Transport Archive (both); [Right] G W Price collection

Next came a high speed run to Prince Alfred Road (known to staff as 'PAR' or 'Prince Alf') so that 544 could be photographed. Another fast run took 972 to Pagemoss via the 49 and 40 then to Pier Head via Prescot Road. From there, it went directly to Kirkby via the 19 returning via the 29 and 49 to Edge Lane and thence via the 40 and the newly-laid track on Brownlow Hill to St George's Place for a refreshment break. From here, the tour went down St John's Lane into Old Haymarket in order to take the 31 tracks to Walton, prior to following the peak hour only 36 to Seaforth. Next came Litherland, from where the final leg took the car along Stanley Road to Lime Street where the tour ended at 7pm. All this for just 5/- (4/- for members).

May 1947: The following cars are known to have been fitted with 'pre-fab' type windscreens during 1946-48.

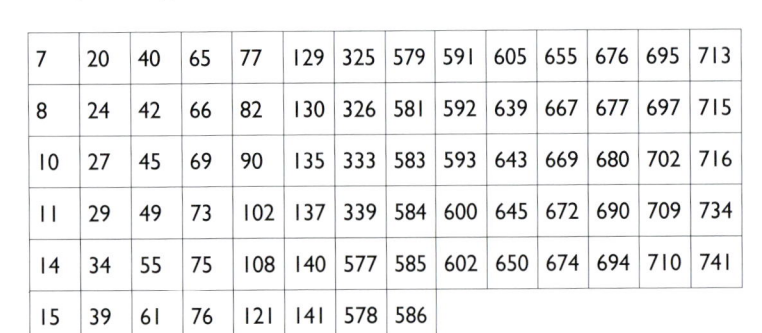

Since the end of the war several cars, including 133, 368, 646, 648, 686, 712 and 717, were fitted with the more robust type of wartime screen.

31 May 1947: "544 on Pier Head-Allerton-Garston circular. Out until 11.30pm [a Saturday] helping to carry the New Brighton trippers home. Normally, this car works only at peak periods on weekdays." (JWG). Jack Gahan took this memorable photo of 544 hard at work on the Aigburth Road reserved track. A family are enjoying the front balcony whilst the driver wears a protective waterproof – just in case. *[Opposite, top] J W Gahan*

21 June 1947: Heading a line of stranded cars on West Derby Road is 163 which appears to have trolley problems. Note the new style of heavier bumper and the louvre ventilators over the upper deck front windows, carried by the later Liners and also the Baby Grands. Behind is Standard No 8. On the same day, Jack Gahan recorded "544 out again all day. 665 (a very good car) repainted from red to green and fitted with windscreens. Balcony car 314 (a travelling ruin) out late Saturday night. Car still in red and cream livery." *[Below] J W Gahan*

7	20	40	65	77	129	325	579	591	605	655	676	695	713
8	24	42	66	82	130	326	581	592	639	667	677	697	715
10	27	45	69	90	135	333	583	593	643	669	680	702	716
11	29	49	73	102	137	339	584	600	645	672	690	709	734
14	34	55	75	108	140	577	585	602	650	674	694	710	741
15	39	61	76	121	141	578	586						

THE LEAVING OF LIVERPOOL

22 June 1947: "Very busy Sunday, thousands travelling 'over the water' (to New Brighton). Many old trams including 544 out all day." **(JWG)**

8 July 1947: After many years, there were sufficient men to begin clearing Edge Lane of life-expired cars. During a visit to the scrap sidings on 26 July 1947, Clifford Noon photographed several doomed veterans including S2 and S3 both of which were original German-built 'Ringbahn-type' passenger cars dating from electrification in 1898. This is S2 which had been converted into a cash van in 1905 and then a depot stores van. It was probably last used sometime in the 1930s. Latterly, it had a Brill truck, Walker 33S motors and S1 controllers. The prefix S is believed to have stood for stores. *[Below] A C Noon/Online Transport Archive*

The operational works fleet was usually housed on the west side of the works. 'Scrubber' PW9 (Permanent Way) had started life in 1901 as an open-top passenger car but had been extensively rebuilt in 1921 and mounted on a 7ft 6in Brill 21E truck. It was one of four cars converted into Scrubbers in 1935. In 1939 it was given the lower saloon from Bellamy 538. *[Above] A C Noon/Online Transport Archive*

23 July 1947: A major power cut brought most of the network to a standstill. Drivers normally coasted to a flat section where the crew could sit it out whilst their unfortunate passengers walked home, as buses were usually full.

28 July 1947: The Transport Committee claimed the Department was losing £5700 a week as costs had increased by 79% since the end of war but revenue by only 63%. A fare rise added a ½d to most of the stages.

A new 6d workmen's return was introduced covering over up to 10½ miles of travel.

Distance (up to)	2 miles	2¾ miles	4 miles	7¼ miles
Fare	2d	2½d	3d	4d

23 August 1947: The loss-making service 16A was withdrawn without replacement.

26 August 1947: On another visit to the scrap area, Clifford Noon noted 432, 580, 582, 589, 590, 595 and 'Rail Scrubber' PW4 (seen here). This was another 1898 former 'Ringbahn-type' motor car fitted with the lower saloon from Bellamy 569 in 1941 (some reports say earlier). A year later, it had sliding doors cut into both sides and, in 1943, windscreens were added. Although marked for scrap in 1944, it remained on site, possibly serving as a store. Formerly S4, it was probably given the prefix PW when the CE&S assumed responsibility for the works fleet in 1943. *[Above] D Conrad*

Jack Gahan recorded many idiosyncrasies: "49 still in red livery has had a continuous 'wobble' for months (bent axle?). Wheels heavily worn and motor noisy. Car returns to service in December painted green and wheels replaced but motors still grinding and noisy." You can almost hear the racket as 49 makes its way along Green Lane where setts are being repositioned around a recently relaid cross-over. Note the complete absence of any barrier or warning signs. Living close to Green Lane, the Gahan brothers monitored happenings on a daily basis. *[Below, left] E A Gahan/Online Transport Archive*

October 1947: The handful of industrial journeys between Penny Lane and Gillmoss or Kirkby assigned the number 48.

12 October 1947: A second LRTL tour this time on 934 which was driven by Inspector Howard. During the 80-mile trip, participants visited Garston, Woolton, Kirkdale, Sheil Road, Bowring Park, Kirkby and Fazakerley. Of particular interest were those sections rarely traversed by the Maleys. Here 934 poses on the long section of single track on Netherfield Road. Note the side bracket arms and imposing residences. *[Below] R B Parr/National Tramway Museum*

42 | THE LEAVING OF LIVERPOOL

14 October 1947: Transport Committee approve the decision to scrap the much-loved Illuminated Car which was sold on 1 December and its remains eventually burnt at the rear of the works in March 1948. It had last been used as part of a Liverpool Hospital Week Appeal in 1939. Some panels were removed for preservation; a small bulkhead panel is at the Wirral Bus and Tram Museum but the whereabouts of others has never been established.

Many were disappointed when it failed to feature in the VE Day celebrations. "People asked why it had been destroyed. Lame excuse was given that there was no room for it and that its scope would be limited as the routes were converted." (JWG) This typified the approach of the Transport Committee. They didn't want anything showing the trams in a good light (pun intended!). Here it as at The Quadrant, opposite Lime Street Station, decorated in all its glory for the Coronation of George VI in 1937. *Martin Jenkins collection/Online Transport Archive*

17 October 1947: Another LTPA public meeting where the vote was again overwhelmingly in favour of tramway retention.

23 October 1947: 908 emerges from the works fitted with its original chrome fenders rather than the post-war more solid steel buffers.

6 November 1947: Did he have a premonition? Jack Gahan opted to photograph Priestly Standard 407 of 1932 as it left North John Street on an early evening working to Green Lane little knowing that a few hours later it would be lost in the worst fire in the undertaking's history, and fires were not uncommon in Liverpool. *J W Gahan*

7 November 1947: About 2.55am fire broke out on Baby Grand 295 parked on track 3. At the time, some 50 workmen, electricians, shunters and cleaners were in the building but most were on their mid-shift break. Conscious of a strong smell of burning, the Fire Brigade was summoned, arriving by 3.05am. 70 firemen were involved. Of the 87 cars (some reports say 88) on shed, 14, mostly from tracks 2 and 4, were driven or pushed to safety on Prescot Road. These included the two vehicles immediately in front of 295. However, an 'obstruction' prevented staff from using the car behind 295 to push it out of harm's way.

1947 | 43

It is probable the 'obstruction' referred to fire hoses laid across the tracks. At 3.30am the Fire Brigade ordered the current to be turned off by which time the roof and wooden cleaning galleries were ablaze although only five or six trams were alight. When the rear section of the corrugated roof collapsed at 4.00am it fell directly onto the trams trapped below. By 5.20am the Fire Brigade had the fire under control although outbreaks continued into mid-morning. Approximately half the building had gone and many local houses had been evacuated. *[Above and previous page, right column]*
Liverpool Fire Service (all)

Marks was woken with news of the disaster – 10% of the tram fleet had been wiped out. "Quite the worst disaster that any tramway undertaking could suffer at the present time ... trams lost irreplaceable under present conditions of work and supply ... an infinitely worse disaster than the damage done by snow last winter when we had as many as 165 trams off in one week." (Transport Department) Marks made urgent appeals to neighbouring bus operators. Amazingly, 26 double-deckers from Wallasey, Birkenhead and Crosville were in action as tram replacements during the morning peak. Others followed from Bolton, Chester, Manchester, Northampton, St Helens, Salford, Southport and Warrington. Included were some 'wrecks' which were quickly returned. Later, 20 more reliable vehicles were purchased from Birmingham and these remained in service for about a year. Also pressed into action were some of the elderly buses purchased earlier from the London Passenger Transport Board. Together with the hired-in vehicles, these covered peak workings mostly on the Prescot Road corridor with some 'expresses' providing a limited stop service from Commutation Row to Finch Lane and beyond. Some 25 luxury coaches were also hired but their reliability proved erratic. A major problem was finding sufficient numbers of trained bus drivers.

The offer of trams from Blackpool and Belfast was turned down due to 'different wheel profiles'. Manchester may also have offered surplus vehicles. At the works, staff made every effort to repair as many laid-up trams as possible. Most of the 750 men employed at Green Lane were transferred to other depots and all Green Lane duties were operated by cars now based at Edge Lane.

By 9am, charred remains were being towed either by breakdown lorries or other trams to the works by way of St Oswald's Street. Among the sorry trio in the lower image are 159 and 912.
[Right top] J B Horne collection; [Right, lower] J W Gahan collection

To avoid over-crowding at Edge Lane some cars, such as 888, went into store at PAR. 295 was kept on site until February 1948 so it could be thoroughly inspected during the subsequent enquiry. Once the front of the depot was declared safe, Prescot Road was re-opened to traffic and tram services restored.

During the rest of the day, any tram that could turn a wheel was pressed into service and deployed across the network so all routes experienced a reduction in service. As a result, many cars appeared in unfamiliar territory often with no route or destination information although a few did have improvised route numbers displayed in the driver's windscreen. "Aftermath of the fire – strange cars, borrowed from other depots, used on Green Lane routes. For instance, 544 actually got to West Derby village on route 12 on several occasions." (JWG)

The Transport Department's response to the fire was rapid, praiseworthy and professional as there was only a minimal loss of capacity during the evening peak.

All known photographs of the fire were taken by the Fire Brigade, Police and local press. To date no views by enthusiasts have surfaced. However, Norman Forbes recorded this scene at Old Haymarket during the evening peak. A pair of Manchester Corporation Crossley buses are loading at the 6A queuing point whilst Standard No 30 is about to leave for Norris Green.
[Opposite] N N Forbes/National Tramway Museum

Amazingly, few people were stranded as they left work early, motorists offered lifts, buses picked up at tram stops and charged tram fares and crews expedited swift turn rounds with one contented passenger commenting "If this is what it does, let's have a fire every day", whilst Councillor Armour told the *Liverpool Echo* the disaster was "a blessing in disguise".

These are the 66 cars destroyed or severely damaged.

13	139	234	292	317	340	391	597*	657	692	876	894	908	960
52	159	256	294	318	367	407	640	665	711	882	895	912	980
74	163	282	295	329	382	420	641	670	713	888	896	915	987
78	173	290	300	334	385	445	649	686	717	892	898	959	991
126	233	291	316	336	386	506*	655	691	728				

* denotes snowplough. Also damaged or destroyed were two salt trailers

884, 885, 886, 887, 906 and 961 spotted in service with visible signs of scorching later in the day. "I saw 906 running in service with badly scorched side upper deck panels – I did not know there had been a fire until the lunch-time." (Ted Gahan/EAG) Some were put out with missing windows covered with cloth or wooden boards.

"We heard the news that there had been a big fire overnight in Green Lane tram sheds. We lingered on the way home hoping to get news how bad it was. Some of the trams that passed were scorched on the bodywork amongst them 886 and 887. There were several others but I have not remembered any numbers. It was some weeks later before I saw my favourite tram 161 which, much to my relief, had not perished. Most of the Green Lane allocation were now 'billeted' on the many sidings around Edge Lane Works." **(Anthony Henry)**

Some of the lost cars are depicted here, others are identified elsewhere in the book. A few days before it was destroyed, 78 is guided into the depot. This view illustrates the confined space immediately in front of the building. *[Overleaf, top] N N Forbes/ National Tramway Museum*

Fire-victim 407 is at Old Haymarket. A split number blind, in this case 'Red' 10C/11, indicated a car returning to depot, in this instance the ill-fated Green Lane. *[Overleaf, centre] E A Gahan*

41, 157, 328 and 338 were eventually repaired and returned to service, the later with a new roof and possibly a truck from a fire-damaged Baby Grand. Standard 41 of 1923 was of special interest. It had been reconditioned before the war with an EMB flexible axle truck, high speed Metro-Vick motors and large style indicators. Shortly after its post-fire repairs, it was involved in a heavy collision in May 1948 and never ran again. *[Overleaf, bottom] J W Gahan*

11 November 1947: 268 was the first car to emerge in the new simplified 'Hall' livery which was described by Ted Gahan as "green with narrow band of cream under top deck windows, lined out in black. Roof and window pillars green – lower parts as before". This view, taken on the North Loop at the Pier Head. The use of the full side lettering was discarded on future repaints, and the side indicators fell into disuse in the early 1950s. *R B Parr/ National Tramway Museum*

11 November 1947: Norman Forbes wrote to the Secretary of the LRTL: "Dear Boylett, The appalling tragedy of Green Lane will probably be a set back to our campaign, but we do not intend to give up the struggle. The Press, mainly anti-tram, is taking the line that the disaster will merely accelerate the conversion so why worry? … Immediately after the disaster

46 | THE LEAVING OF LIVERPOOL

(I have this on fairly good authority) Blackpool was approached re-loan of cars, but it was discovered that the deeper flanges of the Blackpool cars would make them unsuitable for Liverpool track. There is a rumour about some cars being acquired from Manchester but it is no more than a rumour as yet ... of one thing you may be certain; the Transport Committee, headed by Armour, has no intention of acquiring any cars if it can help it."

14 November 1947: Forbes writes again to Boylett: "... the way in which the fire business is being hushed up is distinctly sinister. Armour and Marks have already gone to London to beg priority in the delivery of buses. I may mention, in strict confidence, that at least two towns have offered trams to Liverpool, and they have been refused; and there is evidence that proves the statement made by Marks and others to the Press re the impossibility of obtaining new trams today is not true and that they must know it is not true."

19 November 1947: W G Marks responded to a letter from the LRTL expressing sympathy over the fire: "in view of the fact that Liverpool Corporation have decided on a policy of changeover from trams to buses, my Committee do not intend to purchase either new or secondhand tramway cars to replace those that were lost. They will concentrate more on filling in the gap by purchasing additional buses."

November 1947: Just nine unvestibuled cars remain in passenger service: 43 (Edge Lane); 314, 544, 614, 618 (PAR); 323, 335, 625, 706 (Walton).

December 1947: Another spell of bitterly cold weather with heavy snow ensured all snowploughs were in action.
Official records show there were now 1362 tram drivers divided as follows: Walton 379, Green Lane 299, PAR 180, Dingle 157, Garston 121, Edge Lane 89. Since the fire, Green Lane men still sign on at the depot before going to Edge Lane to pick up their car. Even before the fire, a number of trams covering peak hour duties on the Prescot Road and West Derby Road corridors were housed at Edge Lane but staffed by crews from Green Lane. Time-tables also published the fares charged on cars working back to both depots from these corridors. Officially, any car in Liverpool journeying to and from a depot was 'in service' although some crews were reluctant to obey this rule so they could finish their duty a few minutes earlier.

30 December 1947: The special sub-committee set up to establish the cause of the Green Lane fire released its findings. Questions were raised about the Fire Brigade decision to shut off the power and the absence of a sprinkler system. Councillor John Braddock raised the spectre of the earlier 1942 fire when five trams were burnt. "If improvements had been implemented after the earlier fire it could have been prevented." The failure to install sprinklers in a predominantly wooden structure seems extremely rash. Despite newspaper reports that "one tramcar had jumped the rails as the staff were moving it, thus blocking the only escape route, the destruction of tramcars would not have been so great", the committee declared there was no such derailment. They confirmed "no defects found on 295 when it arrived off duty at 11.55pm" and that the crew had reported nothing unusual. The fire was first discovered by the assistant foreman who smelt burning at 2.57am by which time the whole of the front of 295 (seen below) was ablaze. The committee concluded that the fire started in the resistance cupboard under the stairs and was due to an electrical fault. No reference was made to the fact that cleaning staff sometimes found warmth during their breaks by notching up the controller on a stationary car or by making toast on the resistances. The committee recommended that sprinkler systems should be installed in all premises as well as direct phone lines to the Fire Brigade. For many years, some believed the fire may have been started deliberately especially as the £200,000 insurance money certainly helped to pay for more new buses. *[Previous page] Liverpool Fire Brigade*

Some money was also realised from the sale of three EMB flexible two-axle trucks and some EMB bogies and other equipment to Leeds and six sets of Liner bogies and motors to Glasgow. *[Above] Leo Quinn collection/Online Transport Archive*

Edge Lane also reused some of the salvaged trucks and equipment. During 1948, 815, 829, 836, 851 and 867 were given EMB lightweight trucks, 35hp motors and air track brakes. 853 was also upgraded but without the air track brakes. A mix of lightweight and heavyweight trucks were eventually placed under 879, 881, 889, 957 and 992. For some reason the charred lower deck of 982 survived for several years before being scrapped in May 1950 with its truck going under 181. Freshly outshopped and sporting just the city coat of arms as well as the new lighter font fleet numbers, retrucked 836 is in Carisbrooke Road, Walton.

[Above] Martin Jenkins collection/Online Transport Archive

30 December 1947: The *Liverpool Evening Express* gave full details of the first stage of the conversion programme including the projected order of route closures and associated fare changes, all of which depended upon approval from the Regional Traffic Commissioner. Under Stage One the following were to be converted in this order: 26/27, 23/24, 28/16, 17, 18/18A/36/37, 34, 35, 38, 10, 43/43B, 12, 3, 4/4A/4W/5W, 15, 30/31, 46. In the event, the order was subsequently changed.

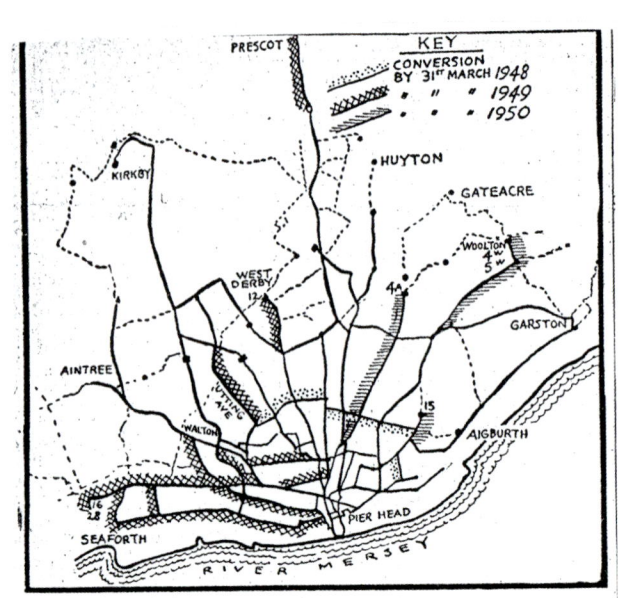

Shaded portions indicate "timing" of scheme to convert Liverpool's passenger transport system from trams to 'buses. See key at top of map.

Green Goddesses Unprepared For Swan-Song?

LAST OF LIVERPOOL'S TRAMS TO RUN UNTIL 1958

L. EXPRESS 30.12.47

26/27 Route Changes To Buses By March

CONVERSION of Liverpool's passenger transport system from trams to 'buses is scheduled to be completed in 10 years.

The scheme will start with the Outer Circular routes No. 26 and 27, and this section is expected to be converted by the end of March next.

31 December 1947: During a New Year's Eve visit, C A Mayou recorded the trams he travelled on: "236 (44), 907 (29), 745 (49), 546 snowplough at Prince Alfred Road, 900 (8A), 958 (33), 5 (16) – Litherland single track in almost SHMD* state of disrepair but Garston track perfect. 790 (Litherland-Stanley Road/Strand Road), 101 (23), 348 (18), 676 (recently vestibuled; Everton Road-Walton), Walton Road track in dreadful state due to dished joints for about a mile. 625 (old livery) at high speed along Netherfield Road to Old Haymarket. 137/577 vestibuled but not painted green. Walked to Lime Street and saw buses some hired from Birmingham on tram routes." (*Transport enthusiasts in the mid-1940s regarded the track at Salford and Bolton and on the last remnant of the Stalybridge, Hyde, Mossley and Dukinfield (SHMD) system, as amongst the worst in the country.)

Some of the cars withdrawn during 1947 were recorded by Jack Gahan. "Cars withdrawn early due to body condition (dates not known) – a few still in red and cream livery – 8, 16, 21, 45, 47, 48, 56, 61 (as 599), 64, 72, 93, 106, 115 (laid up at Dingle for three years), 121, 124, 127, 142, 320, 356, 642, 656, 699, 702, 709, 720, 738, 743. Of these 61, 121 and had just been repainted! Some were accident victims which it was considered uneconomical to repair. Following ran until a defect, usually loose bodywork, leads to withdrawal of 330, 571, 580, 587, 596, 603, 607, 618, 630." Most were broken up but 330 became a cloakroom within the works. The following cars are also believed to have been withdrawn by the end of this year: 58, 306, 432, 578, 582, 665 and 721.

48 | THE LEAVING OF LIVERPOOL

CHAPTER 5 | 1948

Following the decision to scrap the trams Marks, who had been President of the LRTL, felt it appropriate to sever his links. It is of lasting regret that the Corporation failed to give due consideration to his carefully prepared vision for a network of modern light rail lines. The LTPA and LRTL continued to lobby politicians and to organise public meetings but it was clear both political parties were now firmly committed to tramway abandonment. Furthermore, to ensure the policy was a success, many stops were eliminated to improve running times. In reality, the buses rocked and rattled on the poorly-maintained roads. It was during this period that Marks failed to be properly consulted over the transport links needed to serve the burgeoning out of town estates where people were to be rehoused on a first-come, first-served basis, a policy deliberately aimed at breaking the rigid sectarian divide. In July 1948, Marks was succeeded as General Manager by his deputy, W M Hall, a man steeped in the bus side of the industry.

As with the trams, much of the track and surrounding infrastructure was also in poor condition. There had been some rail renewal during the war but, in December 1947, the CE&S claimed that some track was 25 years old and that, more worryingly, 45 miles of single line was veering towards the dangerous. As a result, the CE&S recommended that the Stage One closures should take place as soon as possible but that £1.6 million should be spent on those sections surviving into the later stages. Based on this assessment, the Transport Committee

THE LEAVING OF LIVERPOOL

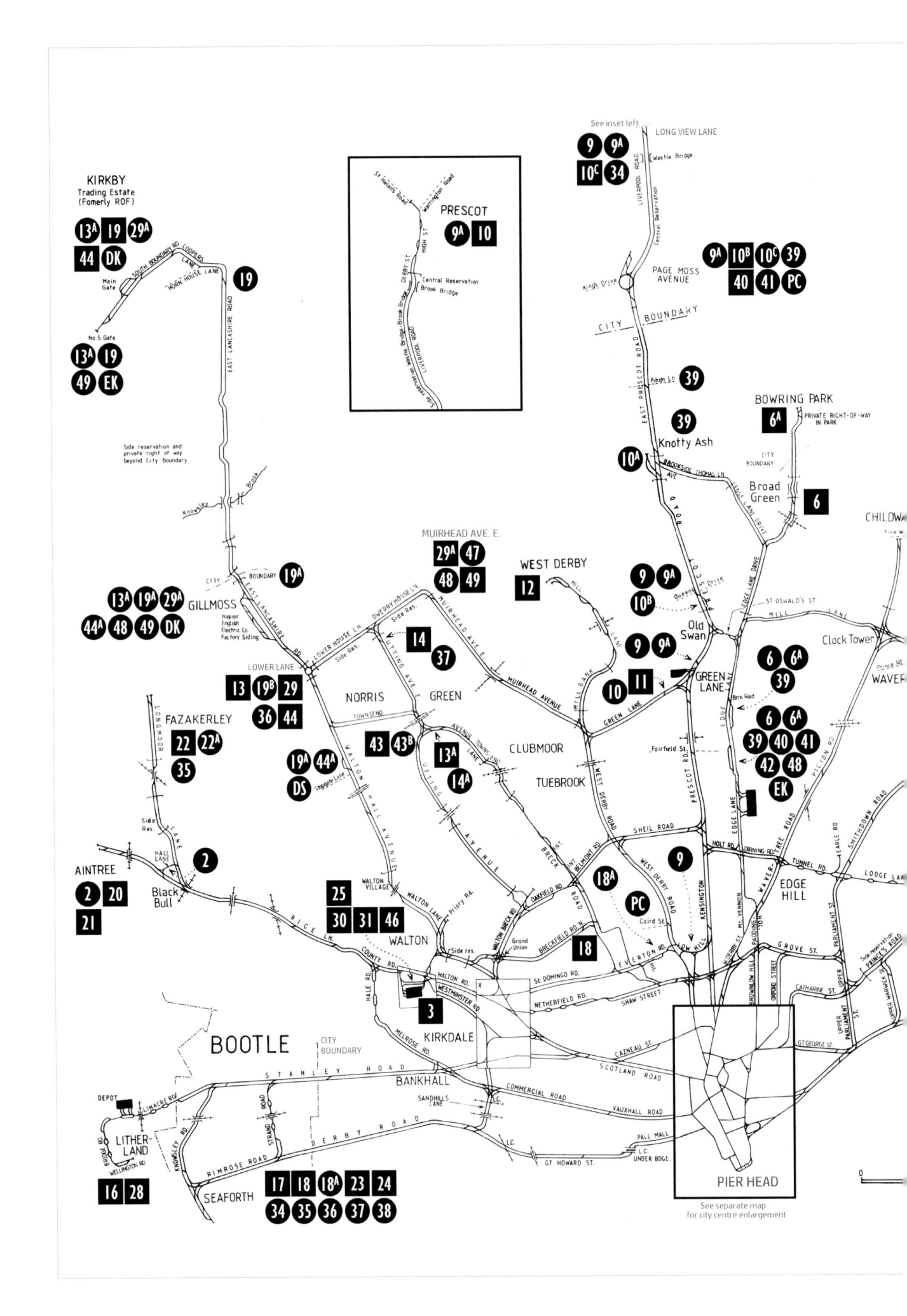

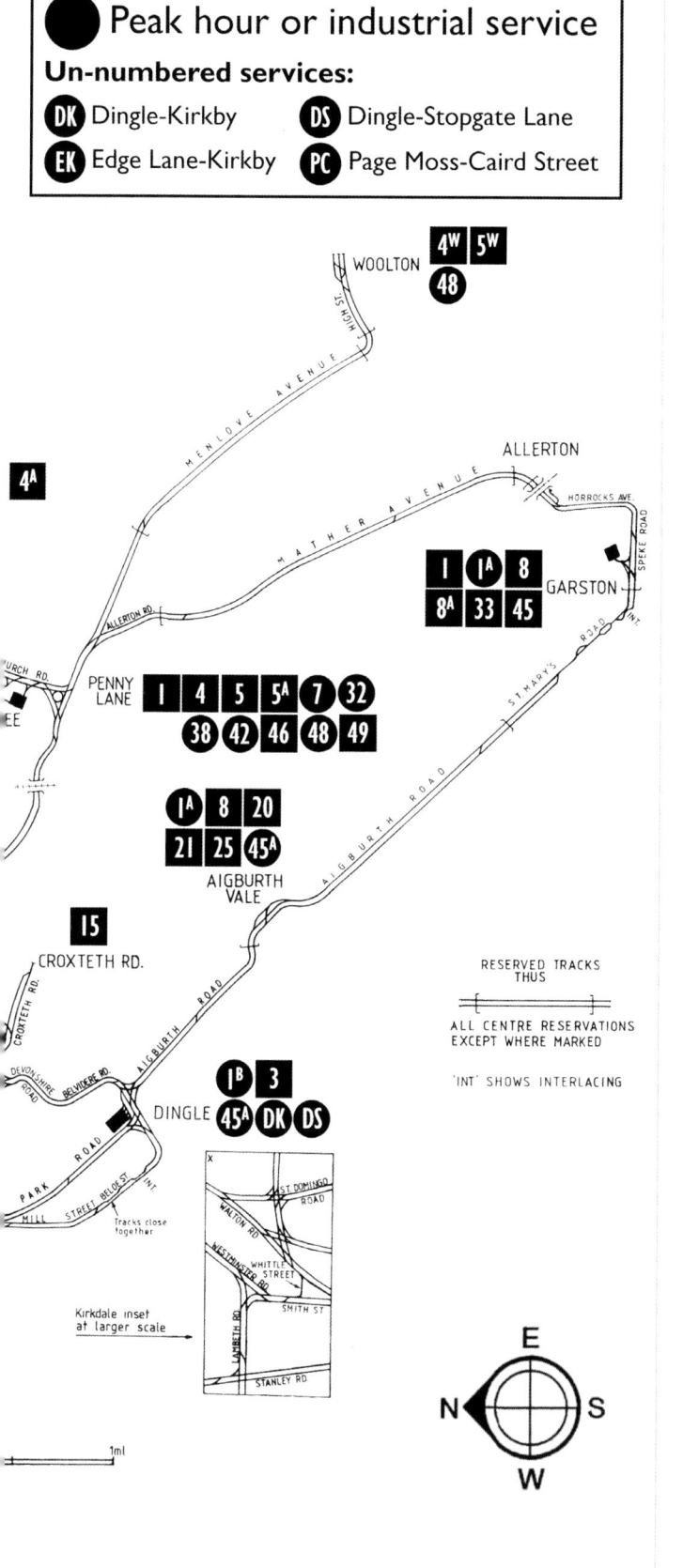

agreed to finance the renewal of some 21 track miles. The LTPA questioned the seemingly excessive cost of renewal when compared to other schemes in cities such as Leeds and Sheffield.

To gain maximum benefit from the new rail, it was agreed the renewals should happen quickly and not be spread over several years. Next the Corporation approached the Ministry of Transport for a loan of £675,000. A ministry inspector was despatched to Liverpool to assess the situation. In his final report, Lt Col Wilson stated "rails in bad condition, some out of gauge especially curves, granite setts loose and breaking up, junctions especially bad, tie bars on reserved tracks irregular, sleepers broken and joints dropped. Most older Standards only fit for scrap." He was concerned about the authorised speed of up to 30mph on reserved tracks. As a result, 15mph restrictions were imposed on Green Lane, Picton Road, Stanley Road, Rimrose Road, Derby Road, Melrose Road and 12mph on Longmoor Lane. At least, he was complimentary about the Liner put at his disposal! In conclusion, Wilson recommended that, for such a large investment, trams should remain until at least 1963. This instantly rang panic bells at Hatton Garden and amongst the anti-tram faction.

In November, the City Engineer produced further facts and figures relating to the conversion. Stage One (1948-52) would involve closing 28.93 miles of which 15.95 miles were overdue for renewal equal to a saving of £986,830; Stage Two (1952-55) 24.75 miles of which 3.8 miles were overdue for renewal at a cost of £223,400 and Stage 3 (1955-58) 35.62 miles of which 7.02 miles needed renewing at a cost of £416,780. At this point, the Kirkby and Garston Circle routes were to survive until 1958. However, at some unknown date, it was decided the 6A/40 would be the last to go as they passed the Works.

Essential track renewals during the year included parts or all of: Kirkdale Road, Walton Road, Walton Village, Walton Hall Avenue (reserved track), Green Lane, Erskine Street, Prescot Road, Crown Street, Pembroke Place, William Brown Street, Byrom Street, Dale Street (Moorfields junction removed), James Street, Renshaw Street, Park Lane, St James Place, Mill Lane (Wavertree), Smithdown Road and Allerton Road (illustrated) as well as the junctions at either end of North John Street, London Road/Prescot Street and London Road/Pembroke Place. *[Page 49] Stan Watkins*

During the clearance of Green Lane, the semi-derelict body of ARP car Bellamy 550 was broken up on site. *[Below] JW Gahan*

Key	Location	All day terminus *Terminal or start point for part-time routes*
A	Pier Head (North Loop)	13 13A 14 16 17 19 19A 22 30 31 43 44 44A 2 14A 43B
B	Pier Head (Centre Loop)	6 6A 10B 10C 29 29A 9 10A
C	Pier Head (South Loop)	1 1A 4 4A 5 5A 8 8A 15 33 45 1B 45A
D	Castle Street	4W 5W 13A 14A
E	Old Haymarket	6 6A 7 10C 13 13A 29A 31 Part of the weekend: 23 24 28
F	Mount Pleasant	5A 8A
G	Commutation Row *	9 10A 10B 10C
H	Roe Street *	14 14A
I	Clayton Square	1A 10A 15 19A 31 39
J	North John Street	11 13A 14A 29A
K	Great Crosshall Street	2 22 43 43B 44 44A
L	Quadrant *	23 24 28 (except parts of the weekend)
M	South Castle Street	10 12 26 27 1A 1B 4A 5A 7 15 26A 27A 32 39 40 45A
N	Whitechapel	20 21
O	Victoria Street	*Football and Races cars*

* Cars for these points usually showed 'Lime Street' as the destination

13 January 1948: Many modern cars again affected by torrential downpours with buses drafted in. "Lots of peak and late evening services cut due to inclement weather." (JWG) This situation continued for several weeks.

26 January 1948: Faced by car shortages and frequent parallel bus routes, many journeys on the 33 were either cancelled or worked by buses.

February 1948: The routes in Bootle should all have gone by now. "More broken rails on Stanley Road than on any other track." (JWG) Since the end of the maintenance agreement between Bootle and Liverpool Corporations in 1942 there had been little or no track work. As relations between the two had not improved, the conversion of the Bootle routes was postponed. However, both sides were now in discussion about ending Liverpool's operating lease and how much compensation Bootle should receive.

March 1948: Other than the Baby Grands, the four-wheel cars In the table opposite are believed to have been on the 'active list' but some were undoubtedly out of service awaiting repair or scrap. Most had two 40hp motors, except for the majority of the reconditioned cars which had two 60hp motors except for 440 which had two 20hp making it the slowest modern tram in the city.

Non-passenger cars, mostly snowploughs, included 122, 134, 303, 502, 507, 513, 527, 539, 540, 546, 553, 555, 565, 566. Balcony 330 used as a 'club house' in the paint shop at the Works and 432, 578, 582 and 721 awaiting disposal.

14 March 1948: 167 cars laid up due to snow damage. "Motors and resistances on modern cars clogged and soaking." (JWG)

19 March 1948: An LTPA protest meeting was held in the Co-op Hall on Lodge Lane to oppose the imminent conversion of the 'Belt' routes: those attending voting 65 to 4 in favour of retention. Concerns were raised about the loss of various concessionary fares and the vital role played by trams during the many 'pea-souper' fogs when visibility was virtually nil and nothing else was moving except the trams. However, the meeting did acknowledge that the post-war boom in demand was abating due to a rise in car ownership. "Retain the trams and keep the cost of living down. Replacement by bus means higher fares and more accidents." (LTPA leaflet)

14-20 March 1948: During Grand National Week, the usual number of extras was run to Aintree at a flat fare of 4d from Pier Head, Victoria Street, Old Haymarket, Rotunda, Carisbrooke Road, Earle Road and other points

52 | THE LEAVING OF LIVERPOOL

Built dates	Type	Numbers
1910	Bellamy, 7ft 6in truck	544
1913-20	Wide-bodied cars, most ex-Double Staircase, 8ft truck	571-573, 577, 579, 581, 583-587, 591-594, 596, 598-603, 605
1914	Ex-Double Staircase with EMB 8ft 6in truck, air brakes	588
1919	English Electric balcony cars with 7ft 6in truck	614, 625, 627, 629, 630
1921/22	Balcony cars, 7ft 6in truck	43, 304, 314, 332, 335, 363, 635, 636
1922-27	Handbrake cars, mostly 7ft 6in trucks	21-27, 29, 30, 58, 69, 70, 72, 86, 110, 117, 133, 311, 325, 326, 333, 348, 358, 376, 378, 608, 634, 734, 736-740, 742
1921-27	Handbrake, 8ft 6in trucks	7-11, 14-18, 20, 29, 34, 36, 39, 40, 42, 45-47, 49, 51, 54, 55, 57, 59-61, 64-68, 73, 75-77, 79, 80, 82, 83, 90, 93, 100, 102, 104, 108, 113, 116, 118, 119, 121, 127, 129-131, 135, 137, 138, 140-143, 308, 312, 321-323, 332, 337, 339, 355, 359, 368, 387, 450, 637-639, 642, 644-648, 650-654, 656, 658-664, 666-669, 671, 672, 674-685, 687-690, 693-706, 708-710, 712, 714-716, 718-720, 735, 739, 741
1925-27	'Long Standards' on EMB 9ft trucks, upholstered seats on lower deck	673, 721-727, 729-732, 745-756
1928-31	Handbrake cars with longer platforms and upper decks, vestibules and upholstered seating on one or both decks, a few transverse in the lower saloon; some fitted with more powerful motors and magnetic track brakes in 1936	1, 5, 6, 12, 62, 63, 71, 84, 85, 89, 94, 107, 109, 112, 115, 120, 123, 124, 139, 144, 146, 149, 301, 302, 305, 307, 309, 313, 315, 319, 320, 327, 329, 331, 344, 356, 359, 372, 373, 383, 389, 400, 410, 415, 429, 432, 441, 462, 572
1936-39	Reconditioned cars with air brakes and EMB 8ft 6in trucks	28, 31, 35, 38, 41, 54, 81, 87-89, 91, 94, 96-98, 101, 105, 114, 128, 147, 305, 334, 338, 353, 386, 393, 420, 445, 451, 454
1932/33	Wide cars, 8ft 6in trucks	111, 310, 324, 380
1935	Wide car with new body and Maley & Taunton 8ft truck	440
1936/37	Reconditioned wide cars with air brakes and EMB 8ft 6in trucks	317, 328, 459, 469
1936/37	Wide cars with reconditioned bodies, air brakes and EMB trucks	342, 343

on the network. Additional cars also operated on routes 20, 21 and 25. Some punters opted to use routes 13, 19A, 29 and 44A as far as Lower Lane from where special buses were run (fixed fare of 1d), whilst others took a 13, 13A or 43 to Utting Avenue-Townsend Lane where more extra buses were provided (fixed fare 3d). According to the day of the week, these extras started any time from noon on 'Jump Sunday' to 10.00am on Grand National Saturday. Cars assigned to these special services usually displayed 'Races' or ran with totally blank screens.

One contemporary report stated that in order to handle the vast crowds, an astonishing 400 cars were needed with some being 'borrowed' from other routes. At the end of each day, thousands made their way home. Published window bills simply said "Return service from Race-Course as Required". In reality, scores of trams struggled to reverse amongst the dense crowds pouring onto the street. Movements were controlled by regulators and inspectors and most cars displayed small cards showing their ultimate destination, eg OH stood for Old Haymarket. Cars departed from Aintree and Warbreck Moor in convoys, whilst cars on routes 20 and 21 terminated at Hall Lane.

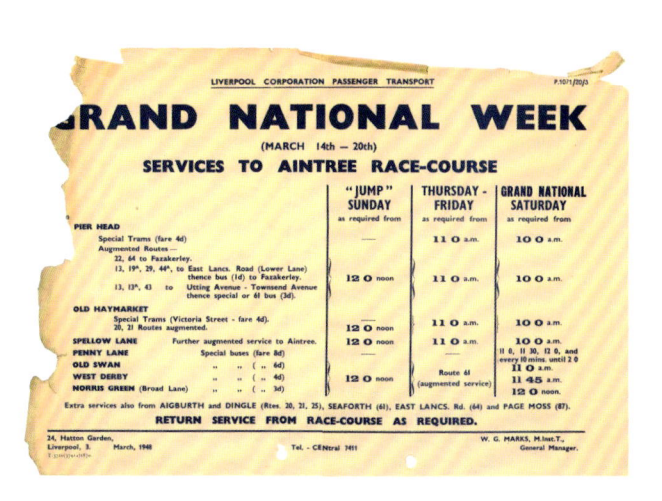

[Both overleaf] N N Forbes/National Tramway Museum; F N T Lloyd-Jones/Online Transport Archive

54 | THE LEAVING OF LIVERPOOL

The Liverpool system had some spurs and sidings used mostly for sporting events. These were at Allerton (South Liverpool FC home games); Walton Breck Road (Liverpool FC home games); Priory Road/Walton Lane (Everton FC home games). The one on Earle Road was used as a boarding point for football supporters going to Anfield and Goodison Park, race-goers bound for Aintree as well as late night audiences from the nearby Pavilion Theatre. It is likely it was last used during Grand National Week although it could have survived until the end of the 1947/48 football season.

April 1948: "If it was possible for vehicles to look any worse in Liverpool, this appeared likely when a strike of body-builders at Edge Lane held up what little maintenance was taking place. About this time visitors to the city asked what colour the trams were meant to be!" (JWG)

14 May 1948: Brake failure caused 954 to run away and derail at the bottom of Kirkdale Road whilst working on route 27. Car was back in service on 27 August having been repaired and repainted. *[Below] Stan Watkins collection*

31 May 1948: The annual LRTL tour was held on Cabin 815 which left its home depot of Walton at 11.20am travelling via route 3 to St George's Crescent, Lime Street where the bulk of the 60 passengers boarded. From there it ran round the Garston circle returning to St George's Place at 1.15pm. From here, 815 followed the 23 to Seaforth from where it travelled to Litherland before going to Prescot via the 34. From there, came a cross-country run via the 10, 49, 29 and 19 to 5 Gate, Kirkby. Following a speedy dash to Walton depot for a refreshment break in the canteen, 815 made for the Pier Head via the 22A, thence to Croxteth Road and Dingle before returning to the city in order to follow the 17 to Seaforth. From here, it covered the 18 and 43 to Norris Green prior to returning to Lime Street via the 14. After most participants had alighted it returned to Walton depot. A remarkable 72 miles for 4/- (non-members 5/-). The driver was Inspector Thomas Howard who is seen alongside the car outside The Imperial Hotel on the peak hour only tracks on St George's Crescent. By this time, 815 had EMB lightweight trucks and more powerful motors salvaged from the Green Lane fire. Despite being overhauled and repainted in 1952 it was withdrawn soon afterwards but not scrapped until 4 March 1953. No tour was ever held with either a Marks Bogie or a traditional Standard.
[Above] R J S Wiseman/National Tramway Museum

Number of standees allowed on the lower deck reduced from eight to five and moves to cover staff shortages led by recruiting more conductresses.

1 June 1948: Fleet now officially stands at 636.

12 June 1948: The first closures involved routes 26/27 which Marks had earmarked for early abandonment in the late 1930s since when virtually no work had been done on the heavily-used track. It had been hoped to replace the routes by 31 March 1948 but the Regional Traffic Commissioner took longer to consider the application for the required road service licences and to consider the Corporation's proposal to abolish concessionary fares with each conversion. These fares included transfers, the popular cheap workmen's returns available before 8am and valid for return at any time of the day and children's school holiday penny returns. The conversion notice warned of another fare change: 'SOUTH CASTLE STREET – the former 'Belt' route and fares THROUGH the Terminus (Victoria Monument) will be discontinued. All journeys will terminate there and passengers may be required to alight.' Until the conversion, people boarding a 'Belt' tram in Scotland Road or Dale Street and could travel through South Castle Street to say Lodge Lane or Holt Road without alighting or paying a second time.

When the road service licences were granted, the all-street track 'Outer Circular' routes 26/26A, 27/27A ran for the last time in the early hours of Sunday 13 June. Worked from Dingle depot and starting from South Castle Street, the 26/26A made the 'north' clockwise run via Castle Street, Dale Street, Byrom Street, Scotland Road, Everton Valley, Walton Breck Road, Oakfield Road, Belmont Road. Sheil Road, Holt Road, Durning Road, Tunnel Road, Lodge Lane, Croxteth Road, Princes Road, Upper Warwick Street, Park Road, Park Lane and Canning Place. Worked from Walton depot, the 27/27A made the 'south' run in the anti-clockwise direction.

Widely known as the 'Belt', they carried substantial traffic through the inner suburbs with many making transfers onto other routes crossed during the eight-mile perambulation. As a result, Dingle crews working the 26 referred to it as the 'Wall of Death', after a popular fairground event of the day. One driver remembered that "after six round trips, you were totally exhausted. It was a devilish road for strap-hangers and short-hoppers." Headways varied between every five and eight minutes to every 10 minutes on Sunday mornings. Whenever possible, larger capacity cars were employed with Priestly bogie cars 770-781 being regulars for many years. During peak hours, some extras came from The Pool at Edge Lane. Until the end, a few cars still displayed red-on-white numbers.

Only a couple of photographs exist of cars showing 26A and 27A which were sometimes shown on cars short working to Rotunda, Sheil Road, Oakfield Road and Upper Warwick Street as well for depot journeys.

Determined to ensure the conversion was a success, 35 buses replaced 30 trams and the overall journey time was cut from 60 to 42 minutes. This 30% time gain was achieved by eliminating 21 (some say 19) of the 60 stops. Overall speed was also improved as the buses no longer had to negotiate sections of single track with passing loops controlled by electric signalling. Some hold-ups had occurred, especially at peak times, when cars ran late due to drivers waiting for passengers changing trams at the numerous transfer points. Some drivers retrained for the buses whilst others moved depots or changed duties. The LTPA drew attention to the buses failure to handle the traffic, highlighting a 43% increase in the number of vehicles and crews required. Overlooked by cranes demolishing the war-damaged Custom House on Sunday 13 June 1948, a replacing AEC Regent II rattles along the setts on South Castle Street. *[Above] Leo Quinn collection/ Online Transport Archive*

With each abandonment, Edge Lane issued lists of cars to be withdrawn or reallocated. Unfortunately, only one of these priceless documents covering conversion of the 13/14 in 1955 has survived. Cars for scrap usually went to Edge Lane to join 'The Pool'. Sometimes, these were used at peak times or for a major sporting event. Changes following a conversion usually took place on a Sunday and were overseen by duty inspectors and depot foremen. On cars definitely going for scrap the following reusable items were removed: lamps, point irons, controller keys, screens (destination/indicator blinds) and, in some cases, seats, panels and controllers, all of which went into store.

Known to crews as 'Belt Cars', 770-781 were reassigned to other routes and some Cabin and Marks Bogies transferred from Walton and Dingle to Edge Lane and Green Lane. "So many modern cars off the road awaiting repair. Indicators are a rarity in the Green Lane area. Streamliners with doors that don't close have to be bound by wire or wedged with the point iron. Cars losing windows and having old sack in their place are left like this for weeks. Cars with side and end panels missing." (C A Mayou)

[Opposite, top left] N N Forbes/National Tramway Museum

Following a conversion, redundant wiring was quickly removed but, in some instances, sections survived as feeders. For example, the wires on part of Durning Road lasted for another nine years! In the case of the track, the CE&S would often lay new through rail especially at worn out junctions, although often the severed remains were left in situ.

12 June 1948: As the replacing buses for route 26 were worked initially from Edge Lane, space was urgently needed. As a result, five tracks at Green Lane came back into use with 166 being the first to enter. Eventually, all 10 tracks were reactivated. Despite this, extras for the Prescot Road and West Derby Road corridors came from The Pool at Edge Lane. *[Below] A D Packer*

13 June 1948: As a result of the 26/27 conversion, the handful of morning journeys on dockers' route 34 (Longview Lane-Seaforth) now made a lengthy detour between Old Swan and Everton Valley by way of St Oswald's Street, Edge Lane, Paddington, Crown Street, Moss Street, Brunswick Road, Everton Road and St Domingo Road. Evening trips from Seaforth were not quite so long. Having reached Everton Road, cars used Low Hill to reach Kensington ready for the long haul to Longview Lane.

23 June 1948: Disaster averted when 762 was discovered smouldering in Dingle depot. Fortunately, the flames were quickly extinguished. Today, this is one of the cars operating on the Wirral Heritage Tramway.

July 1948: On an unknown date, the following cars were listed at Dingle as awaiting attention at the works: 8 (both ends smashed), 115 (out of use for years), 123, 215, 255, 587 (laid up for many years), 666, 683, 720, 931 (platform bearings collapsed), 932, 990.

10 July 1948: Marks succeeded as General Manager by W M Hall. Shortly after his appointment, local enthusiast, Stan Watkins sent him a letter in which he raised a number of points some of them relating to individual trams. For example, he asked why 884 had been at the back of Garston depot for five years together with 253 which had drooping ends. Although the original letter does not survive, Stan retained Hall's reply which is dated 9 December 1948. Here are some of his responses. "The general

1948 | 57

ROUTE PROFILE
26, 26A, 27, 27A ('The Belt')

Scotland Road. *Martin Jenkins collection/Online Transport Archive*

Walton Breck Road. *J H Roberts/Online Transport Archive*

Belmont Road/West Derby Road. *J H Roberts/Online Transport Archive*

Sheil Road – only known view of a 26A. *E A Gahan*

Tunnel Road. Negotiating one of two passing loops. *N N Forbes/National Tramway Museum*

58 | THE LEAVING OF LIVERPOOL

Lodge Lane, a narrow road with three passing loops. *N N Forbes/National Museum*

Lodge Lane/Croxteth Road. *J H Roberts/Online Transport Archive Tramway Museum*

Upper Warwick Street/Park Lane. *J W Gahan*

Journey time	60 minutes
Last day of operation	12 June 1948
Cars seen in service on the last day	968, 986
Last cars	986 (26) and 968 (27) left South Castle Street at midnight and arrived back at their depots shortly before 1.00am on Sunday 13 June 1948.
Track abandoned	Part of Walton Breck Road, Oakfield Road, Belmont Road, Sheil Road, Holt Road, Durning Road, Tunnel Road, Earle Road siding, Lodge Lane, Upper Warwick Street. Many worn out junctions with new straight through track.

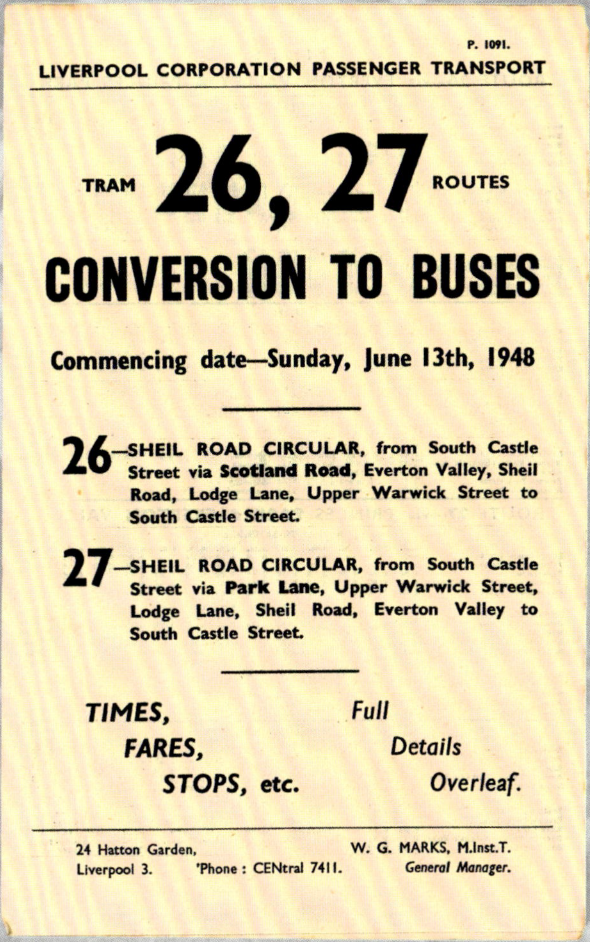

P. 1091.

LIVERPOOL CORPORATION PASSENGER TRANSPORT

TRAM **26, 27** ROUTES

CONVERSION TO BUSES

Commencing date—Sunday, June 13th, 1948

26 —SHEIL ROAD CIRCULAR, from South Castle Street via **Scotland Road**, Everton Valley, Sheil Road, Lodge Lane, Upper Warwick Street to South Castle Street.

27 —SHEIL ROAD CIRCULAR, from South Castle Street via **Park Lane**, Upper Warwick Street, Lodge Lane, Sheil Road, Everton Valley to South Castle Street.

TIMES, FARES, STOPS, etc.   Full Details Overleaf.

24 Hatton Garden, Liverpool 3.   'Phone : CENtral 7411.   W. G. MARKS, M.Inst.T. General Manager.

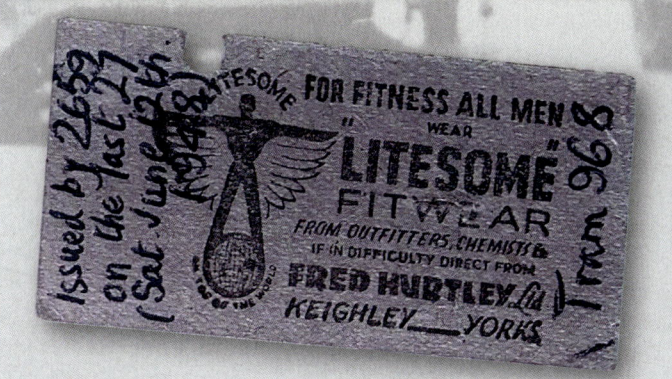

1948 | 59

condition of the bodywork and paintwork of the tramcars is governed entirely by the present labour force, and although a considerable amount of work is now being undertaken and output has increased, both from the Body and Paint Shops, it will be some time before the fleet regains anything like its pre-war condition, but I would like to assure you it is not intended that the tramcars be allowed to deteriorate until they reach scrap condition as you seem to suggest. The reason that certain of the newer type of tramcars have been held up awaiting repair whilst older types have continued to run, has been entirely due to the labour position and the vehicle shortage, for instance, cars 884, 877 and 932 were all extensively damaged, in addition to bodywork, required mechanical and electrical attention and priority had to be given in the Works to repairs which were not so extensive to keep the maximum number of vehicles in service."

13 August 1948: Freshly outshopped 868 jumps the points at the Crown Street/West Derby Street junction and collides with 988. Eleven slightly injured. Both cars back in service later in the year.

7 October 1948: Liverpool's last new tram completed. Owned and built by the CE&S in their workshop at Breckside Park, CE&S 234 was transferred by road to Edge Lane. Its truck, motors, controllers and platforms were derived from withdrawn Standards but all its internal parts were new including the motor-driven grinding equipment which was lowered and raised to and from rail level. Painted in two shades of grey and numbered in the CE&S fleet it was in regular nocturnal use grinding or scrubbing the significant levels of rail corrugation which was such a noisy feature of the post-war system. *[Right - top, centre] Liverpool City Engineers & Surveyors (both)*

29 November 1948: Laid up at Litherland: 31, 123, 204, 218, 278, 288, 323, 601, 653, 704, 705, 708, 730, 742. Together with 704, 31 was broken up on site. The latter had almost certainly suffered terminal damage when it was blown off the tracks when the SS *Malakand* exploded in Huskisson dock on 3 May 1941.

11 December 1948: The first cross-city route to be replaced was the 3, which linked Dingle to Walton via Park Road, Great George Street, Renshaw Street, Lime Street, London Road, Norton Street, St Anne Street, Cazneau Street, Scotland Road, Kirkdale Road, Smith Street, Westminster Road and Barlow Lane. This 4¾ mile all street track route had a peak demand of eight cars with headways mostly every 15-20 minutes but increasing to every 11 minutes on Saturday afternoons and decreasing to every half hour Monday-Friday evenings. Operated from Dingle depot, it was known by crews as 'Wezzie' or 'The West' as it travelled along Westminster Road. The only known short working was southbound to Mill Street.

This abandonment led to closure of the first major sections of track in the central area involving three-quarters of a mile of badly-worn rail on Norton, St Anne and Cazneau Streets used by routes 3, 24 and 28. Trams no longer served these narrow streets fringing the central area with their mix of light industry, warehouses, shops and older houses. Included was an awkward section of single track and loops plus a seriously dilapidated junction with Islington.

This closure led to routes 23, 24 and 28, which had terminated at The Quadrant (seen here), being rerouted to South Castle Street. The Quadrant terminal tracks were abandoned but St John's Lane (foreground) and St George's Place (right) were retained for peak hour and emergency use only. This led to a further loss of traffic as the trams no longer served the main shopping areas. To make space at South Castle Street, routes 39 and 40 were extended to Pier Head although some peak hours extras still used South Castle Street. *[Opposite, top] Martin Jenkins collection/Online Transport Archive*

12 December 1948: Dingle based cars known to have been reallocated: 35, 38, 950 and 986 to Green Lane and 786, 811, 848 and 856 to Walton.

December 1948: A new type of wind-up windscreen fitted to the Liners starting with 158, 954, 170, 160, 988 and 868.

December 1948: Determined to be better prepared for the winter, ex-Double Staircase car 594 fitted with ploughs but scrapped in early 1949.

7/9 December 1948: The sorry fate of English Electric cars 758, 763 and 768 highlighted the Transport Department's failure to capitalise on its assets. The trio had lain dormant in Dingle since 1936 because they had experimental control equipment and inside frame trucks which made them prone to derailing. To alleviate the chronic car shortage, they should have been equipped with trucks and equipment salvaged from the Green Lane fire. Instead, they were passed for scrap. 763 was still in its original red livery whilst 758 and 768 were in pre-war green. On 7 December, 758 was towed to Edge Lane by 608 followed two days later by 763 and 768 each of which was towed by No 11. All went via Garston, Penny Lane, Church Road and Edge Lane. Purchased by George Cohen they were quickly broken up. "A shameful episode." (Eric Broadbent) *[Right] Stan Watkins*

1948: The following cars are understood to have been withdrawn during the year and sold to local scrap dealers Maden & McKee located on Prescot Road. As in previous years, the precise date on which individual cars last ran in service was not logged nor the exact date on which they were broken up and burnt. When a car was identified for scrap this was recorded in the Transport Committee minutes as was the day it was sold to the scrapman who then decided when it should be broken up. In some instances, the actual month of withdrawal is known as shown below. However, a few Pool cars were occasionally laid up for some time possibly as a source for spares.

May	596, 621, 708
July	113
August	8
September	666
November	21, 31, 41 (collision), 72, 83, 93, 100, 117, 120, 142, 146, 320, 387, 578, 599-601, 607, 614, 627, 704, 735, 738
December	656

Other cars known to have been disposed of: 45, 47 (lower saloon sold to St Helen's Girl Guides), 48, 54 (state of collapse 29 May 1948), 64, 121, 124, 323, 356, 454, 550 (long-withdrawn Bellamy), 587, 601, 618, 629, 630, 644, 653, 705 and 730.

ROUTE PROFILE
3 ('Wezzie')

Dingle. *N N Forbes/National Tramway Museum*

Dingle. *F E J Ward/Online Transport Archive*

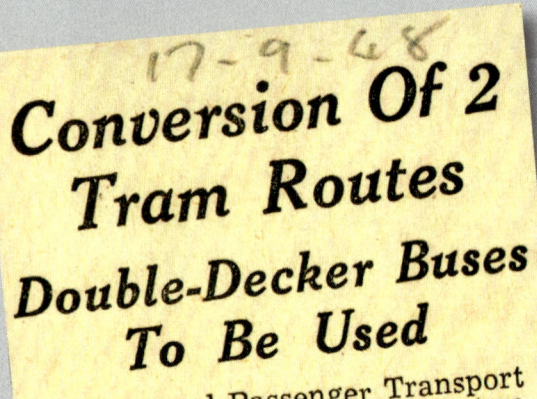

17-9-48

Conversion Of 2 Tram Routes

Double-Decker Buses To Be Used

Liverpool Passenger Transport are preparing to convert two more tram routes to buses, on the No. 3, Dingle to Walton (Spellow Lane), and No. 43, Utting Avenue (Broad Lane) to Pier Head or Great Crosshall Street, services.

Cazneau Street, passing boarded up North Market. *N N Forbes/National Tramway Museum*

62 | THE LEAVING OF LIVERPOOL

St James Place, showing tower of Anglican Cathedral under construction in background. *N N Forbes/National Tramway Museum*

Lime Street. *N N Forbes/National Tramway Museum*

St Anne Street/Islington. *N N Forbes/National Tramway Museum*

Journey time	**38 minutes**
Last day of operation	11 December 1948
Cars seen in service on the last day	761, 769
Last cars	Northbound: 761 (11.20pm Walton to Dingle). Southbound: not known.
Track abandoned	The Quadrant, Norton Street, St Anne Street, Cazneau Street, Westminster Road, Barlow Lane

1948 | 63

ROUTE PROFILE
3 ('Wezzie') (continued)

LIVERPOOL CORPORATION PASSENGER TRANSPORT

P. 1136 (Con. 2).

TRAM ROUTE

DINGLE 3 WALTON

CONVERSION TO BUSES

Commencing date—
Sunday, 12th December, 1948

ROUTE—via Park Road, Park Place, St. James Place, Great George Street, Berry Street, Renshaw Street, Lime Street, London Road, Norton Street, St. Anne Street, Cazneau Street, Scotland Road, Kirkdale Road, Smith Street, Westminster Road, Barlow Lane.

TIMES, Full
FARES, Details
STOPS, etc Overleaf.

24 Hatton Garden, Liverpool 3.   'Phone CENtral 7411.   W. M. HALL, General Manager

St Anne Street, near to St Anne's church. *N N Forbes/National Tramway Museum*

Westminster Road. *N N Forbes/National Tramway Museum*

Barlow Lane. *N N Forbes/National Tramway Museum*

64 | THE LEAVING OF LIVERPOOL

CHAPTER 6 | 1949

Routes now tumbling thick and fast with each closure releasing more elderly handbrake cars for scrap although too many newer cars still off the road gathering dust. *[Above] N N Forbes/National Tramway Museum*

Winter snow and slush continues to flood motors on the Liners – a problem which occurred with new trams delivered to Aberdeen which also rode on EMB trucks.

The massive MOT loan financed a track renewal programme spread over two years and included key sections of the Garston Circle including Smithdown Road, Allerton Road, Aigburth Road, Belvidere Road and Princes Road all of which were relaid and the roads resurfaced. "Miles of new tracks laid knowing it would all be ripped up in a few years' time. Chronic waste." (C A Mayou)

The LTPA continued to make the case for retention. Norman Forbes and others attended public hearings. They hoped the MOT recommendation to delay final closure until at least 1963 might bear fruit but, not surprisingly, the Transport Committee won the day. They argued that any delay would increase the overall cost of the conversion by over £1¼ million. Furthermore, the annual accounts seemed to 'prove' the trams had lost a massive £384,000 during the financial year 1948/49. These figures were challenged by the LPTA but not by the press or the full Council. Following the MOT intervention, some members of the LTPA believed Hatton Garden had adopted an unofficial policy of reduced tramway maintenance in order to press the case for speedy abandonment. The chairman of the LTPA was now local solicitor H E C Piercey and the Secretary C H Postance.

Due to the vehicle shortage, no attempt was made to assign newer, more powerful cars to dedicated routes. As a result, older, slower, underpowered cars led to bunching especially in rush hours. The trams were also losing passengers to parallel bus services especially after protective fares charged on outbound off-peak buses were discontinued.

January 1949: 707 and 708 fitted with snowploughs but never used.

15 January 1949: The state of the rail on the 1½ miles of reserved track along Utting Avenue led to replacement of routes 43/43B. Speed restrictions of 15 and 20mph had been enforced since the Ministry inspection, with boards warning drivers. "Riding a 43 was akin to a switch-back railway. In places no check rail and joints lost in a sea of mud and filthy water." (C A Mayou) Worked from Walton depot, the 43 was a busy, trunk route with headways varying between six and 10 minutes. It linked Pier Head to Utting Avenue via Georges Dock Gate, Water Street, Dale Street, Byrom Street, Scotland Road, Everton Valley, Walton Breck Road, Wylva Road/Arkles Road and Utting Avenue.

43B was used for evening peak departures from Great Crosshall Street, depot journeys and other short workings. One photograph shows a 43B with Norris Green/East Lancashire Road on its screen which meant it would have gone beyond its usual terminus at Utting Avenue in order to reach Lower Lane via either the 13 or 14. *[Below] E A Gahan*

ROUTE PROFILE
43, 43B

LIVERPOOL CORPORATION PASSENGER TRANSPORT

TRAM ROUTE

UTTING AVE. **43** CITY

CONVERSION TO BUSES

Commencing date—
Sunday, 16th January, 1949

ROUTE—via Utting Avenue, Arkles Road (Wylva Road outwards), Walton Breck Road, Everton Valley, Kirkdale Road, Scotland Road, Byrom Street, Dale Street, Water Street, Pier Head.
(Additional peak hour Terminus at Gt. Crosshall Street)

Scotland Road/Bevington Bush. Note the loading islands, the ornate lamp standards and the fully-loaded car on what was known locally as 'Scottie Road'. *F N T Lloyd-Jones/Online Transport Archive*

Wylva Road. *N N Forbes/National Tramway Museum*

Utting Avenue/Bootle Branch railway bridge. *E A Gahan*

Utting Avenue/Queens Drive. *E A Gahan*

Journey time	33 minutes
Last day of operation	15 January 1949
Cars seen in service on the last day	956
Last cars	956
Track abandoned	Utting Avenue continued to be used by industrial service 37, post-match football extras and cars on the 13A and 14A working back to depot from Broad Lane although, as from 17 January 1949, drivers were meant to take the longer way back via Walton Hall Avenue.

66 | THE LEAVING OF LIVERPOOL

29 January 1949: When Liverpool FC played at home, extras, often showing 43, departed from Pier Head and Victoria Street. Having off-loaded, some returned to Walton depot whilst others lined up along Walton Breck Road from where post-match specials departed for Lime Street and Clayton Square, whilst others loaded in Wylva Road for Utting Avenue East. Despite the conversion of the 43s, trams still worked these football specials for over a year. Car 380 of 1933, the last Priestly Standard to be built, is making its way onto the siding. Note the complicated tram layout and the small OH in the window for 'Old Haymarket'. *[Above} N N Forbes/National Tramway Museum*

On the same day, car 10 was withdrawn for scrap when it was in collision with Corporation bus L452 at the junction of Rocky Lane and Lower Breck Road, 14 people being taken to hospital.

22-26 March 1949: Open balcony survivors 43, 544 and 625 among cars used during the Aintree Spring meeting culminating with Saturday's Grand National.

1 April 1949: Aimed at speeding up tram travel in the south end, new Ultimate ticket machines were introduced at Dingle and Garston depots issuing 1d, 1½d, 2½d, 3d and 4d tickets.

14 May 1949: Route 15 (Pier Head-Croxteth Road) was the shortest on the system and was duplicated throughout by frequent bus services. It had closed before the war but was quickly reopened on 11 September 1939. Linking the central area to more prosperous southern suburbs, it operated via James Street, Church Street, Ranelagh Street, Renshaw Street, Leece Street, Catherine Street, Princes Road and Croxteth Road. Four cars provided the off-peak 10 minute headway which dropped to 20 minutes in the evenings and on Sundays. Just over half a mile of worn rail was abandoned. After the war, 15A was assigned to morning peak hour extras to Clayton Square and South Castle Street and to evening departures from the latter. No known photographs of cars showing this number.

21 May 1949: The first West Derby Road route to be abandoned was the 12, the outer section of which had been considered for closure before the war. It linked South Castle Street to West Derby via Church Street, Lime Street, London Road, Moss Street, Brunswick Road/Erskine Street, Rocky Lane, West Derby Road, Mill Bank and Mill Lane. Closure had been set for 31 March 1949 but was delayed due to objections raised by local residents. The track between the junction with Muirhead Avenue and the stub terminus in West Derby Village was in shocking condition. There were frequent derailments and a 15mph speed restriction was in force on the half-mile of reservation on Mill Bank. "The bolts holding the rails together had as much as a ¼ inch play in them and sank 4½ inches into the ground when a car ran over them." (D Gratton). Cars also picked their way cautiously in and out of the two loops on Mill Lane. Worked by Green Lane depot (with extras from Edge Lane if required), the 12 ran an eight or 10 minute service except on Sunday mornings when it was every 15 minutes. Normally, 12 cars sufficed whereas 16 buses were needed to cover the extension to Cantril Farm. The abandonment ended the practice of morning peak hour cars from West Derby to North John Street showing 11 and westbound cars from Muirhead Avenue showing 12 when terminating at either North John Street or South Castle Street by way of Lime Street and Church Street. Just over a mile of track was involved.

23 May 1949: Towards the start of the morning peak, the air brakes on 223 failed as it descended St Domingo Road causing it to hit Standard 27 which toppled over as it came down Everton Valley. Some Corporation workmen assisted the injured to the pavement, most having cuts from flying glass. Nine ambulances were called. At the time, priority was given to reopening the road as swiftly as possible and reinstating the tram service.
[Below} Stan Watkins collection

May 1949: Surprisingly, handbrake Standard 701 appeared in a new livery featuring a cream band below the upper deck windows. It is possible the paint shop mistook it for a reconditioned air-brake Standard! Various cars laid up in PAR – 270, 242, 272, 342, 874 plus accident victims 469 (out of service since 1948), 751, 916, 933 and 924 with low tyres.

ROUTE PROFILE
15, 15A

Dingle depot. *A C Noon/Online Transport Archive*

Journey time	23 minutes
Last day of operation	14 May 1949
Cars seen in service on the last day	695, 759, 784, 848, 849, 990
Last cars	849 left Pier Head at 11pm. Handful of enthusiasts on board who, on reaching Croxteth Road, stayed on for the run to Dingle depot.
Track abandoned	Croxteth Road

Croxteth Road terminus. Last day. In Liverpool, cars returning to depot showed split route numbers. *A S Clayton/Online Transport Archive*

Routes 15, 15A.
CROXTETH ROAD—CITY

INWARD STAGE No. TO CITY		OUTWARD STAGE No. FROM CITY
1½d. FARES (WORKMEN'S RETURN 2d.)		
1	Croxteth Road and Upper Parliament Street	3
2	North Hill Street and Leece Street (Berry Street)	2
3	Leece Street (Berry Street) and Pier Head	1
2d. FARE (WORKMEN'S RETURN 3d.)		
1	Upper Parliament Street and Pier Head	1
2d. FARE		
1	Croxteth Road and Pier Head	1

Princes Park Gates. *E A Gahan*

68 | THE LEAVING OF LIVERPOOL

Princes Park Gates. This worn triangular junction was replaced by straight-through curves, the outward track being relocated within a new roundabout. *A S Clayton/Online Transport Archive*

Pier Head, South Loop. *Leo Quinn collection/Online Transport Archive*

1949 | 69

ROUTE PROFILE
12

West Derby village. *J H Roberts/Online Transport Archive*

Mill Lane, West Derby station. *J W Gahan*

Mill Bank. *E A Gahan*

Issued by 100 on the last Route 12 tram — Sat May 21st 1949. Tram 338.

Islington Square. *E A Gahan*

LIVERPOOL CORPORATION PASSENGER TRANSPORT

P. 1175 (Con. 5)

TRAM ROUTE

WEST DERBY **12** **SOUTH CASTLE ST.**

CONVERSION TO BUSES
AND EXTENSION TO CANTRIL FARM

Commencing date—
Sunday, 22nd May, 1949

Journey time	33 minutes
Last day of operation	21 May 1949
Cars seen in service on the last day	32, 112, 338, 341, 459, 650, 667, 673, 690, 800, 807, 885, 899, 906
Last cars	338. Made the last full round trip finally leaving South Castle Street at 11.45pm with a fair number of enthusiasts on board. Quite a crowd at West Derby to wave the car off as it leaves for Green Lane depot at 12.20 on the Sunday morning. For part of this run, 338 driven 'unofficially' by LRTL member Ted Gahan.
Track abandoned	Millbank, Mill Lane. Wires swiftly cut, sleeper track lifted and the former junction with Muirhead Avenue relaid with new curves.

West Derby Road. *E A Gahan*

1949 | 71

3 June 1949: S W Dale noted the following at the rear of the Works: 7, 32 ("body dangerously loose"), 55, 66, 70, 76, 107, 111 ("very loose body, used for towing and shunting until withdrawn February 1949"), 122, (snowplough), 134 (snowplough), 143, 149, 303 (snowplough/works car), 311, 315 ("bad condition"), 319, 332, 372 ("very loose"), 454 ("laid up since 1948"), 513 (snowplough), 592, 683, 721, 727, 739. Most soon met a fiery end but 592, 727 and 739 were returned to service.

4 June 1949: Maley 924 re-enters service with the trucks off 933.

25 June 1949: The next route to be replaced was the long 'interurban' 10, known officially as 'The Prescot Light Railway'. Worked from Green Lane (with extras from Edge Lane), it was the first route using the Prescot Road corridor to be replaced. From South Castle Street, cars travelled via Church Street, Clayton Square, Lime Street, Kensington, Prescot Road, East Prescot Road, Liverpool Road, Derby Street and High Street, Prescot to the awkward single track terminus in the middle of busy St Helens Road where conductors had to avoid the traffic and the trolleybus wires when swinging the trolley. Following representations by Norman Forbes to the North-West Traffic Commissioner, the original conversion date was postponed. However, all objections were overruled. Prescot Council highlighted the dangerous state of the single track and loops along the narrow parts of Derby Street and High Street as well as the cramped terminal spur which they claimed caused congestion. Approximately 1½ miles was abandoned. It

> "As a young boy I went with my mother on a 10. I had read about trolleybuses in Prescot and wanted to see one. I remember the conductor was astonished when we told him why we had come. I was fascinated by the trolleybuses and we had a ride to St Helens and back. I still remember our return run from Prescot. The Standard was making heavy weather with lots of grating and crunching sounds. When we got to Green Lane, we were all turfed off and told to wait for the next car. Another memory – my mother had a friend who lived near West Derby. We would go on the 12 or more likely a 10 and walk across the fields. I recall one wartime journey I think on a Cabin, which had wooden boards in place of some missing upper deck windows. This seemed to me to make the upper deck very dark and mysterious." **(MJ)**

is believed that this also marked the end of the Knotty Ash stub. Until the loss of the 10, it had been possible to travel under wires (tram/trolleybus) from Pier Head to Atherton, Leigh, Swinton and Bolton.

The 10s ran every 15 minutes weekday mornings but every 20 during the rest of the day whilst on Sundays it was every 10 minutes.

Catering for those shopping and working in Old Swan, additional cars ran between Green Lane and Prescot (sometimes Longview Lane) every 10 minutes Monday to Saturday. They departed from a crossover at the south end of Green Lane (seen here) which was also terminus for route 11 but in the opposite direction. In the morning peak, extra 10s also worked to Old Swan, Low Hill, Commutation Row (Lime Street on indicators), Old Haymarket, Clayton Square and North John Street. In the evenings, extras departed from all but the last two places. To help shift the enormous evening crowds, express, minimum fare buses also operated between the City and Prescot. *[Bottom left] GW Price collection*

Since 1947, a peak hour tram returning from Prescot, Longview Lane or Pagemoss was meant to show 9 if bound for the city and 9A if only going to either Green Lane or Edge Lane depot. Following conversion of the 10, the use of 9A gradually fell into disuse. One of the Priestly Lengthened Standards is seen on Prescot High Street. The triangle on the dash indicated the car was equipped with air brakes. *[Above] N N Forbes/National Tramway Museum*

25 June 1949: Last unvestibuled cars believed to have been withdrawn en bloc, including 43, 314, 335, 544, 625, 627, 635 and 706. Some reports suggest withdrawal may have taken place on an unknown date in July. *[Opposite, top] A W V Mace/National Tramway Museum*

The above were magnets for photographers especially 544, the last survivor of a one-time fleet of 445 cars with Bellamy roofs. Dating from 1910 it had a Brill-type 7ft 6in 21E truck, two 35hp motors, BTH18 controllers and two and two transverse seating in the lower saloon. Following a body inspection at the Works, it was returned to passenger duties in 1942 still in its pre-war, red and cream livery. Very much PAR's 'mascot', the 68-seater worked almost the same duties for several years. In the mornings, it was a return trip to South Castle Street usually as a 32 but occasionally a 4A or 7 whilst, in the evenings, it was a single return trip from Penny Lane to and from Napier siding at Gillmoss as a 48. Here, it is outside the Carlton cinema on West Derby Road on 1 July 1948. At weekends and on bank holidays, it could be found shifting crowds returning by ferry from New Brighton or else working football and races specials. *[Opposite, bottom] E A Gahan/Online Transport Archive*

72 | THE LEAVING OF LIVERPOOL

"The sight of 544 inevitably caused astonishment among people in the streets not to mention passengers who thought it came out of the Ark! Upon its appearance drivers of cars going in the opposite direction would make mysterious gestures, stamp on their gongs or kow-tow. Enthusiasts seek out 544 at every opportunity, one even going so far as to desert his girlfriend in Church Street on seeing 544 and making a mad dash for it." **(JWG)**

ROUTE PROFILE
9A, 10

St Helens Road, Prescot. *All images: F N T Lloyd-Jones/Online Transport Archive (unless where noted)*

High Street Prescot, peak hour 9A returning to Green Lane depot. *N N Forbes/National Tramway Museum*

74 | THE LEAVING OF LIVERPOOL

High Street, Prescot

Derby Street, Prescot

Liverpool Road.

Pagemoss.

LIVERPOOL CORPORATION PASSENGER TRANSPORT

P. 1173 (Con. 6)

TRAM ROUTE

PRESCOT **10** SOUTH CASTLE ST.

CONVERSION TO BUSES

Commencing date—
Sunday, 26th June, 1949

Issued by 4727 on the last 10 Saturday 25th June 1949. Tram 144

Journey time	50 minutes
Last day of operation	25 June 1949
Cars seen in service on the last day	144
Last cars	144
Track abandoned	St Helens Road, High Street, Derby Street and Liverpool Road reservation. Although the rail dated from the 1920s the three-quarter mile of roadside reservation was in reasonable shape.

1949 | 75

> "I recall being virtually ordered by my elder brothers to go for a last ride as 544 was due along Green Lane shortly on route 48. The order was obeyed, and a ride on the upper deck back balcony enjoyed as far as Penny Lane. Quite an adventurous trip for an unaccompanied 8 year old." **(Anthony Gahan/AFG)**

Following its official withdrawal, the 39 year old veteran was used for driver training but, all too soon, it was broken up. "Attempts to have 544 preserved were not successful; how could they be when the council and transport department had become violently anti-tram? The excuse was they had no room for museum pieces." (JWG)

26 June 1949: Although both entrances remained, the tracks in Green Lane depot were reduced to the five on the east side giving a capacity for around 40 trams. This was to make way for the growing number of buses.

24 July 1949: Annual LRTL tour held on English Electric bogie car 769 with Inspector Howard as driver. Starting from its base at Dingle, the car picked up the 50 or so participants in St George's Place, outside the Imperial Hotel. Departing at 11.30am it ran first to Childwall Five Ways and Woolton from where it returned to Wavertree in order to reach Longview Lane via the 49 and 10C after which it ran to Pier Head via route 40.

Following a break outside the Imperial, the next leg took 769 round the Garston Circle to Dingle where there was a quick tour of the depot followed by a run to Litherland via routes 1 and 28, where the car is seen in Bridge Street with Norman Forbes on the platform. 769 then proceeded to Seaforth from where it followed the 17 to the Pier Head. After a break for tea, the tour proceeded to Kirkby via the peak hour only track along Utting Avenue. The final leg took it back to Dingle depot via routes 44, 36, 16 and 45. Despite some fast running, its controller didn't overheat once! *[Above] J H Roberts/Online Transport Archive*

August 1949: 342 is moving slowly along new rail on Green Lane. Note the stacks of setts on the left. This was one of the 60-seat cars built at Edge Lane in 1932 but later reconditioned with 8ft 6in EMB flexible axle trucks, airbrakes and more powerful 60hp motors. *[Below] E A Gahan/Online Transport Archive*

76 | THE LEAVING OF LIVERPOOL

12 August 1949: Peak hour 33s from South Castle Street and Clayton Square discontinued.

13 August 1949: The life-expired track on Park Road led to the loss, without replacement, of the 1 and 20 as well as a major south-end route reorganisation. Just over 1½ route miles abandoned.

On leaving the Pier Head (South Loop), the 1 reached Garston via James Street, Church Street, Ranelagh Street, Renshaw Street, Great George Street, Park Road and Aigburth Road. At some point before Garston, conductors changed the screen to show 8 so, on arrival, the car returned to the city by way of Penny Lane. If only going as far as Aigburth or Garston, cars usually showed 1A. The combined headway on the 1/1A was 12 minutes Monday to Friday, 20 minutes for most of Sunday and every six minutes on Saturday. The 1B was peak hours only. However, as with so many Liverpool peak hour extras there was no hard and fast rule. For example, morning extras working in from either Garston, Aigburth or Dingle to either Clayton Square, South Castle Street or Pier Head might show 1, 1A or 1B whilst evening extras from Pier Head, South Castle Street or Clayton Square could show – but not always – 1A when terminating at Aigburth or Garston or 1B for Dingle or Aigburth. Although worked from Garston depot, some early morning and late night duties may have been provided by Dingle.

Dingle was the base for the busy cross-city route 20 Aigburth to Aintree via Aigburth Road, Park Road, Paradise Street, Whitechapel, Old Haymarket, Byrom Street, Scotland Road, Walton Road, County Road, Rice Lane and Warbreck Moor. Reflecting the high volume of traffic especially on the northern section, headways varied between six to 10 minutes Monday to Saturday and 10 minutes all day Sunday.

Short workings to such points as Walton Church, Black Bull (Hall Lane), Walton (Spellow Lane) and Dingle were officially designated 20A. To help alleviate over-crowding on the southern leg, there was an evening queue point at the north end of Whitechapel which it is believed was accessed by extras from Dingle depot by way of Lime Street, St George's Place and St John's Lane. Following home games at Goodison Park, Everton fans boarded cars from special queue points on County Road (northbound) and Nimrod Street (southbound).

14 August 1949: Following the loss of the 1, frequency on the 8 was increased throughout the week to between five and 10 minutes, on the 33 to every 10 minutes except Saturdays when it increased to every six minutes, but on the 45 the headway was reduced to half hourly throughout the week. Together with the 8 and 45, the 33 now formed part of the revised Garston Circular. The 45 was also rerouted via Park Lane, Canning Place and South Castle Street from where cars to Pier Head (South Loop) travelled via James Street, but from Pier Head they left via Water Street and Castle Street. *[Above] J H Roberts/Online Transport Archive*

To compensate for loss of the 20 and frequency reduction on the 45, the cross-city 21 was rerouted via Mill Street and Beloe Street and scheduled to run every eight minutes except on Sundays when it dropped to every 10 minutes.

1 September 1949: The driver of 393, Albert Wildman, has a 'miraculous escape' following a collision.

14 October 1949: S W Dale made further observations at Edge Lane: 23, 39, 46 ("in shocking condition – it had been withdrawn in May 1948 but reinstated until withdrawn again September 1949 – complete ruin"), 54 ("state of collapse – withdrawn 29 May 1948"), 60, 71 ("travelling ruin"), 79, 80, 108, 122, 134, 303, 315 ("bad condition – withdrawn August 1949"), 502, 507, 513, 540, 546, 553, 653, 654, 689, 726.

ROUTE PROFILE
1, 1A, 1B, 20, 20A

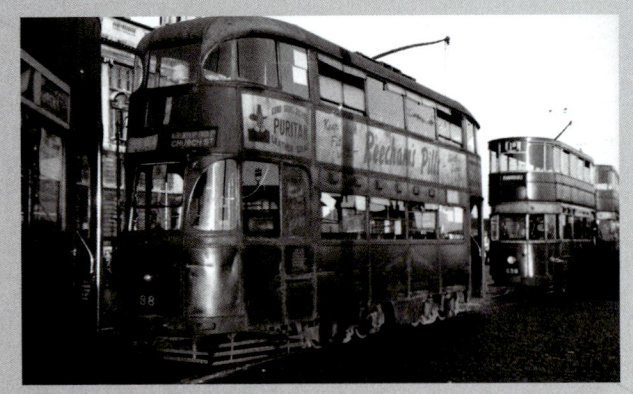

Pier Head, 1946 siding. Last day view. The two-line via displays on indicators were dispensed with during the early 1950s. *A S Clayton/Online Transport Archive*

Berry Street, with St Luke's Church on the right, retained as the "Bombed-out Church". *Photographer unknown/Online Transport Archive*

Park Road. *N N Forbes/National Tramway Museum*

Park Road, passing Dingle Liverpool Overhead Railway station. *N N Forbes/National Tramway Museum*

Dingle. *N N Forbes/National Tramway Museum*

Aigburth Road/Woodlands Road. *H B Christiansen*

Penny Lane. *Stan Watkins*

Aintree. *F N T Lloyd-Jones/Online Transport Archive*

78 | THE LEAVING OF LIVERPOOL

Rice Lane. *I Davidson*

Walton. *R B Parr/National Tramway Museum*

St James's Street, depot working *N N Forbes/National Tramway Museum*

Park Road. *E A Gahan/Online Transport Archive*

Park Road. Last night view. *A S Clayton/Online Transport Archive*

Journey time	41 minutes (1), 54 minutes (20)
Last day of operation	13 August 1949
Cars seen in service on the last day	154, 638, 849, 880, 970, 990
Last cars	154 (1), 849 (20, Aintree-Aigburth), 970 (20, Aigburth-Aintree)
Track abandoned	Berry Street, Great George Street, part of St James Place, Park Place, Park Road, St George's Place

LIVERPOOL CORPORATION PASSENGER TRANSPORT
P. 1181 (Con. 7).

GARSTON – CITY
SERVICE ALTERATIONS

TRAM ROUTES

1, 20 — Discontinued

8, 21, 33, 45 — Services revised

BUS ROUTES

82 — Re-routed via Park Road

82D — New supplementary service between Dingle and City

Commencing date—
Sunday, August 14th, 1949

1949 | 79

14 October 1949 was also the last day for peak hour route 5A which ran to and from Penny Lane via Smithdown Road. In the morning, 5As ran to Clayton Square or South Castle Street and in the evenings to Penny Lane from South Castle Street and Renshaw Street or Mount Pleasant, the latter leaving town via Oxford Street and Grove Street instead of Leece Street. On 14 May 1948, former Double Staircase car 600 of 1919 is seen at Derby Square. For a while this had been experimentally mounted on bogies from one of the early American-built single-deckers. It was withdrawn in November 1948. *[Above] E A Gahan*

On the same date, the 48 operated for the last time to and from Woolton although there may have been some peak workings during the Saturday morning peak but no records survive.

15 October 1949: This closure, which involved the loss of nearly 2½ route miles, marked the beginning of the ruthless cull of the south end services operated from PAR. Although the recently relaid Smithdown Road section of the interlinked 'Belt' routes 4 and 5 was meant to outlive the Wavertree Road section, the decision was taken to replace both these all-street-track routes. From the Pier Head, 4s reached Penny Lane via Dale Street, London Road, Pembroke Place, West Derby Street (outbound)/Paddington, Crown Street (inbound), Picton Road, Wavertree Road and Church Road. On arrival at Penny Lane, blinds were changed for return to the Pier Head as a 5 by way of Smithdown Road, Upper Parliament Street, Catherine Street, Leece Street, Renshaw Street, Ranelagh Street, Church Street, Lord Street and James Street. The 5s worked the 'Belt' in the opposite direction changing to 4s at Penny Lane.

Their suffix gave the 4W/5W a whiff of exotic mystery. These Woolton routes were the only ones to terminate on Castle Street. However, the constant stream of through trams in both directions meant there was no layover time for the crew. On departure, 4Ws duplicated the 4s and 5Ws the 5s as far as Penny Lane from where they proceeded to Woolton via Allerton Road, Menlove Avenue and High Street, Woolton. During school holidays, hundreds of children from across the city took advantage of the scholar's penny return to reach Calderstones Park, Camp Hill and Woolton Woods.

The condition of the grass track along Menlove Avenue and High Street, Woolton hastened their demise. "Speed not always possible on the two miles of grass tracks. Some stretches, because of poor wartime maintenance, were in bad condition.

Thus rail joints were often exposed and flapped up and down as trams passed over them. If too much speed was attempted, the cars fairly rocked and rolled along the uncertain track." (J Amey)

The Corporation claimed to have saved £216,000 on track renewal. However, the planned replacement of routes 4A, 7 and 32 had to be postponed due to a shortage of buses and/or crews. On 15 October, two-thirds of a mile of track on Mount Pleasant and Oxford Street used by routes 8 and 8A was also abandoned together with the peak hour loading point at the city end of Mount Pleasant, which until the Friday evening had also been used by some 5As. 762 is ascending Mount Pleasant in October 1949. Eric Vaughan recalled "until 15 October, the last 8 of the evening went south via Renshaw Street instead of Mount Pleasant". *[Left] N N Forbes/National Tramway Museum*

The outbound track on Ranelagh Street used by routes 4A, 5, 5A, 5W, 8, 8A, 33, 39 and 40 was also abandoned. This view, taken early in 1949, shows car 946 en route to Croxteth Road.
[Above] E A Gahan/Online Transport Archive

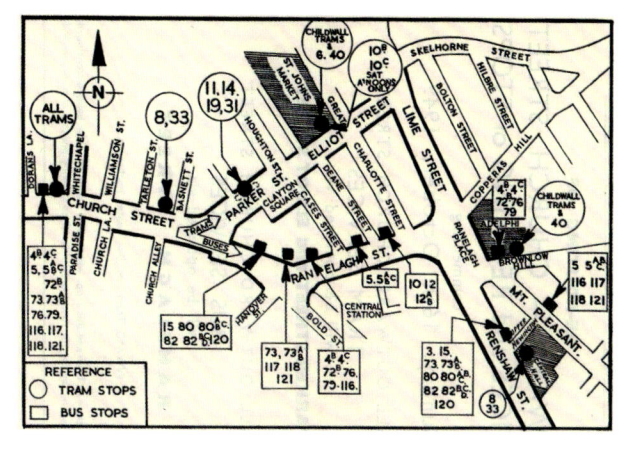

16 October 1949: All outbound cars using Church Street now routed via Parker Street and Elliot Street. This leaflet detailed these revisions as well as the re-siting of stops.

ROUTE PROFILE
4, 4W, 5, 5W

Castle Street. *A S Clayton/Online Transport Archive*

Lord Street. *Photographer unknown/Online Transport Archive*

Leece Street. *N N Forbes/National Tramway Museum*

Upper Parliament Street/Catharine Street, Rialto junction. *N N Forbes/National Tramway Museum*

Penny Lane. *Stan Watkins*

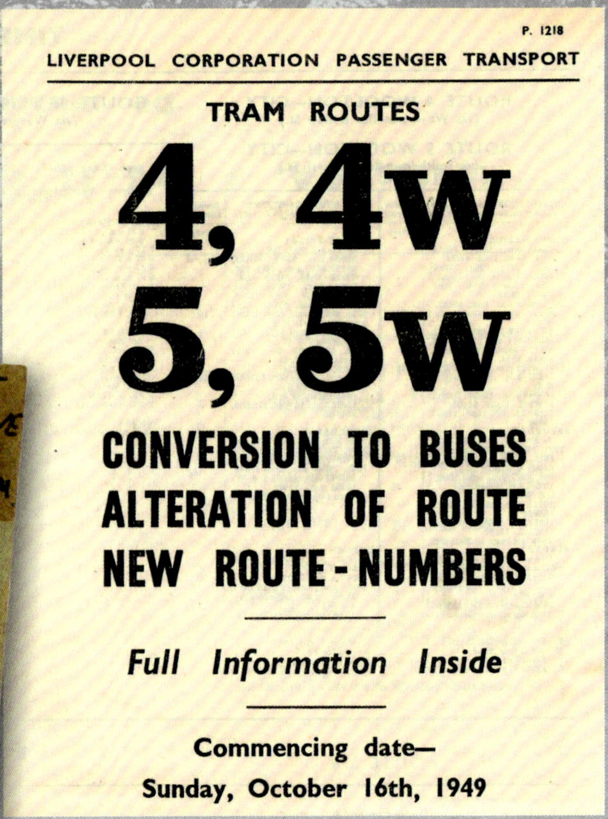

P. 1218

LIVERPOOL CORPORATION PASSENGER TRANSPORT

TRAM ROUTES

4, 4w
5, 5w

**CONVERSION TO BUSES
ALTERATION OF ROUTE
NEW ROUTE - NUMBERS**

Full Information Inside

Commencing date—
Sunday, October 16th, 1949

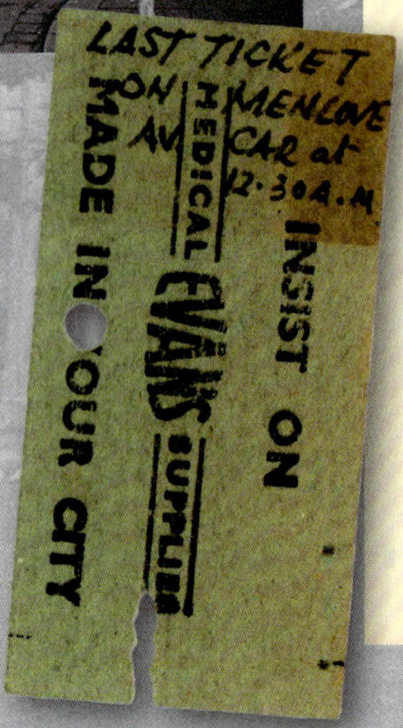

82 | THE LEAVING OF LIVERPOOL

Menlove Avenue. *N N Forbes/National Tramway Museum*

Allerton Road, Woolton. *F N T Lloyd-Jones/Online Transport Archive*

High Street, Woolton. *N N Forbes/National Tramway Museum*

Woolton. *A S Clayton/Online Transport Archive*

Journey time	30 minutes (4/5), 41/42 minutes (4W/5W)
Last day of operation	15 October 1949
Cars seen in service on the last day	102, 242, 247, 251, 276, 331, 342, 358, 368, 440, 645, 678, 791, 869, 874, 875, 878, 911, 920, 922, 923, 927, 936, 958
Last cars	920 (4 City-Penny Lane), 923 (5 City-Penny Lane), 874 (last 4W round trip), 869 (last 5W round trip). Well-loaded with enthusiasts and watched by local residents, 869 finally left Woolton at 12.25 on Sunday morning 16 October 1949.
Track abandoned	Menlove Avenue and High Street, Woolton

Although no longer displaying a route number, the 4A continued as a 'Temporary Revised Tram Service' with the majority of journeys via Brownlow Hill and Church Street but supplemented by a few using London Road and Dale Street. Speed restrictions remained in force on Wavertree Road.

Following these abandonments, there was a major fleet reallocation, details of which were recorded by Ted Gahan and Stan Watkins. The object was to reduce maintenance costs by assigning certain groups of cars to given depots, something the LPTA had been advocating for several years.

PAR to Dingle	440, 918-942 (although some remained at PAR awaiting rebuilding at Edge Lane)
Dingle to Green Lane	772, 776, 944, 946, 988, 990
Dingle to Edge Lane	759, 767, 769
Dingle to Walton	778, 780, 784, 788, 789, 825, 826, 834, 836, 842, 847, 849, 852, 853, 860, 867
PAR to Garston	765, 766
PAR to Edge Lane (30.10.49)	760, 764

17 October 1949: Route 48 curtailed to run from Penny Lane to Gillmoss or Kirkby and the 8A evening queue point relocated to Renshaw Street.

5 November 1949: Peak hour routes 7 and 32 replaced. Due to the shortages of cars and staff, the former had no published time-table, with trams sometimes working a morning journey from Penny Lane to South Castle Street via Church Road, Wavertree Road, Paddington, Crown Street, Boundary Place, London Road and Dale Street. These were balanced by occasional evening departures from South Castle Street or Old Haymarket.

In contrast, the 32 was well-used by dockers using workmen's returns to access the south end docks. From Penny Lane, cars travelled via Smithdown Road, Upper Parliament Street, St James Place, Park Lane and Canning Place. During Monday to Saturday peaks, the 32 ran approximately every 10 minutes. In the evenings, cars were usually full leaving South Castle Street.

After the morning rush, many crews returning from the Pier Head and South Castle Street took the more direct 32 tracks back to PAR. Another 10 trams released for scrap.

7 November 1949: Another visit by S W Dale to the scrap sidings: 46, 60, 71, 73, 322, 393, 644 ("in state of complete ruin, virtually fell to pieces in service, withdrawn October 1948, laid up and scrapped 1950"), 651 and 654.

At the Parker Street/Church Street junction there was a serious collision between 810 on the 6A and 974 on the 33 resulting in 13 passengers being injured.

15 November 1949: Choking fog was a recurring hazard. At times, visibility was so bad nothing moved except the trams which were often followed by a line of cars and bicycles. Despite being on fixed rails, accidents did occur especially on the reservations. At about 7.30am, 903, 774, 184, 857, 741 and 698 were involved in three different but connected shunts within 100 yards of each other on fog-bound Walton Hall Avenue. Filled with workers for Littlewoods Pools and Kirkby Trading Estate, all six were on the same track. Liner 903 lost air pressure and slid on the wet rails probably into 698 which was at the Abingdon Road stop. A third car, possibly 741, was hit in the rear seven or eight minutes later whilst it waited for the wreckage ahead to be cleared. Then two others collided a little further back. Despite the severity, everything was on the move again within 45 minutes with three of the cars continuing in service whilst the others, including 903 and 774 (seen here), were towed away to Edge Lane to be repaired. *[Top right] Martin Jenkins collection/Online Transport Archive*

21 November 1949: Routes 30 and 31 extended from Walton to Aintree to provide much-needed additional capacity on this trunk corridor following withdrawal of route 20. Again, the replacing buses had failed to cope. However, Liverpool Corporation were determined to retain this presence in Ribble territory so any future tram to bus conversion would be granted a licence. 685 is outbound on Lime Street shortly after the routes were extended. *[Above] R Dudley Caton*

11 December 1949: The cull of the Wavertree Road routes was completed with replacement of the former 4A and the loss of a further 2½ miles of track. The route linked Childwall

84 | THE LEAVING OF LIVERPOOL

Five Ways to the Pier Head via Childwall Road, Picton Road, Wavertree Road, Paddington (inbound), Crown Street/West Derby Street/Mount Vernon/Irvine Street (outbound), Brownlow Hill, Ranelagh Street (inbound), Parker Street/Elliot Street/Lime Street (outbound), Church Street, Lord Street, James Street. Worked from PAR, trams passed close to the loco shed and massive marshalling yard at Edge Hill with its rows of terraced houses mostly occupied by railway families. Cars ran every 10 minutes Monday to Saturday and every 15 minutes on Sundays. Short workings included depot runs to and from PAR. A few cars may still have turned at South Castle Street. There were also some journeys via London Road and Dale Street with short workings to and from Old Haymarket. Although always associated with the South Loop, there is a view of a 4A on the Centre Loop! Most cars showed "Childwall/ Five Ways" or "Childwall".

Following cars believed to have been reconditioned during 1949: 260, 168, 174, 887, 966. *[Above] Leo Quinn collection/Online Transport Archive*

Cars believed to have been withdrawn and sold to Maden & McKee during 1949. As in previous years, the precise date on which cars last ran in service is often not known nor the exact date on which they were broken up and burnt. After withdrawal, some were laid up for quite some time possibly as a source for spares.

January	17, 42, 53 (snowplough), 119, 123, (laid up in Litherland for two years), 323 (snowplough), 584, 586, 643, 720, 742 ('the Crystal Palace' – the first all enclosed car)
February	111 (fast and powerful and as such ended up as Works tow car)
March	96, 311, 332, 721, 740
April	9, 66, 107 (fast), 143, 454
May	27 (accident), 120 (withdrawn November 1948), 581
June	43, 314, 335, 544, 625, 635, 706
July	7, 32 (body dangerously loose), 55, 149, 372, 661, 736, 756 (accident, no roof)
August	23, 79, 80, 108, 304, 315, 599, 653
September	39, 40, 46, 58, 122 (snowplough), 579
November	60 (fast), 71 (very fast), 73, 393, 471, 651

Also withdrawn during 1949: 10, 42, 68, 70, 76, 119, 134 (former passenger car), 301, 303 (former passenger car), 317, 319, 322, 348 (lower saloon sold to Huyton Council), 469, 504 (snowplough), 654, 661, 683, 689, 705 (featured in the film *Waterfront*), 715 and 726.

ROUTE PROFILE
7, 32

Old Haymarket. A rare view of a car (ex-Double Staircase) on route 7. *N N Forbes/National Tramway Museum*

South Castle Street. *A C Noon/Online Transport Archive*

Upper Parliament Street, Rathbone Street. *N N Forbes/National Tramway Museum*

86 | THE LEAVING OF LIVERPOOL

Upper Parliament Street, Hope Street. *N N Forbes/National Tramway Museum*

Penny Lane. *Stan Watkins*

LIVERPOOL CORPORATION PASSENGER TRANSPORT

P. 1230

TRAM **7, 32** ROUTES

CONVERSION TO BUSES

(with Route 7 increased to all-day service on Weekdays)

Commencing Monday, November 7th, 1949

Journey time	28 minutes (7), 26 minutes (32)
Last day of operation	5 November 1949
Last cars	Not known (7), 102 (32)
Track abandoned	Part of Upper Parliament Street

1949 | 87

ROUTE PROFILE
4A

Journey time	30 minutes
Last day of operation	11 December 1949
Cars seen in service on last day	160, 276, 754, 875
Last cars	Two different reports – either 160 or 875
Track abandoned	Childwall Road (two-thirds of a mile of centre reservation), Picton Road, Wavertree Road and Irvine Street

Childwall Five Ways. Leo Quinn collection/Online Transport Archive

LIVERPOOL CORPORATION PASSENGER TRANSPORT
P. 1224

CHILDWALL TRAM ROUTE DISCONTINUED

BUS **79** ROUTE

REVISED SERVICE

(Alternately via Brownlow Hill—Church St., 79, and London Road—Dale Street, 79A)

Commencing Monday, December 12th, 1949

Picton Road, inbound from Wavertree Clock Tower. N N Forbes/National Tramway Museum

Wavertree Road, crossing the railway bridge. N N Forbes/National Tramway Museum

Brownlow Hill. F N T Lloyd-Jones/Online Transport Archive

88 | THE LEAVING OF LIVERPOOL

CHAPTER 7 | 1950

Stage One of the conversion programme well underway. No real improvement at Edge Lane Works with an appalling lack of much-needed heavy maintenance. Things were so bad that W A Schofield, Secretary of the Liverpool Engineers and Craftsmens Guild alleged there was "no organisation and ample mismanagement". The LPTA may well have had an inside source, as they obtained very detailed weekly breakdowns of the number of cars off the road some of which were still out of service for months. Some of these statistics are detailed below:

6 May 1950	80 in works and 84 in depots. Total 164 plus nine accident victims in the works and 32 in various depots. Cars in works for over 6 months: 921, 963, 973, 976, 978, 982 (Green Lane fire victim), 984, 986.
3 July 1950	105 in works.
18 July 1950	103 in works.
27 July 1950	110 in works and 80 in depots, including 30 Baby Grands and 37 Liners.
1 August 1950	115 in works. 921, 973, 976, 978, 984 now in for over 8 months.
15 September 1950	108 in works including 20 Baby Grands and 34 Liners.

After Hall submitted a report on the state of the fleet, the City Council sanctioned a significant rehabilitation programme involving 146 Liners and Baby Grands some of which like 253 had lain at Garston from 1944 until September 1950.

Major track renewal was also taking place but only on those routes the MOT advised should then survive until 1963. However, when Cllr Armour led a delegation to the Ministry in London in June, he said the public were no longer prepared to accept the losses incurred by the trams. As a result, the MOT agreed to a final closure date of 1958.

January 1950: "New green interior livery first applied to 750 followed by 802, 887, 891, 877, 299, 246, 884, 793 and 762. Only 137, 577, 694 and 709 remain in red and cream." (JWG)

6 February 1950: To reduce the risk of accidents caused by cars crossing the busy main road, the attractive Bowring Park terminus with its two crossovers and Bundy time-clock was abandoned after service on the Monday night. *[Top right] F N T Lloyd-Jones/ Online Transport Archive*

7 February 1950: From the first car on Tuesday, cars on the 6/6A terminate at the end of the reservation on Broadgreen Road.

February 1950: On some unknown date, the one-way line on Wylva Road and the north track on Utting Avenue were abandoned and lifted. Cars on route 37 now used Arkles Road and the surviving southern track on Utting Avenue in both directions.

March 1950: To provide direct access from Muirhead Avenue and West Derby Road to the central shopping areas, some inbound 29As rerouted to North John Street via Lime Street and Church Street.

25 March 1950: Augmented service on routes 21, 22, 30 and 31 to cater for the Grand National crowds. Some Races specials are known to have operated from Pier Head and Old Haymarket but not in the same numbers as in previous years. 605 of 1920 approaches the railway bridge spanning Warbreck Moor. *[Overleaf, top] N N Forbes/National Tramway Museum*

Jack Gahan recorded the cars involved in this Aintree line-up: 84, 129, 677, 67, 76, 339, 99 and 326. Others known to have carried race-goers on Grand National day include 605, 608, 690, 762, 887 and 950. In all, 77 cars left Aintree and Warbreck Moor, often in convoy, conveying 6400 passengers to the city between 4.30pm and 5.40pm. *[Overleaf, centre] J W Gahan*

Dingle to Garston	918-942, 951
Litherland to Garston	203, 279, 287
Edge Lane to Garston	207, 267
Walton to Garston	183, 184, 903, 947, 955, 962, 966, 968
PAR to Garston	242, 251, 273, 283, 868, 869, 872-875, 903, 968

The Baby Grands proved ill-suited to duties on the fast moving Garston Circle as their elderly controllers tended to overheat if worked for too long in the same direction so most were transferred to other depots, except for 242, 269 and 271 which were normally confined to the 8A or 45, which enabled drivers to change ends at Garston after each journey.

1-3 April 1950: Changes to allocations at the three south end depots, however these exchanges recorded by Stan Watkins and Ted Gahan may not be a complete record. Also, in the early 1950s, cars could be loaned to other depots usually to cover for the 146 Liners and Baby Grands being reconditioned.

Garston: Allocation of 58 cars for off-peak turnout of 36 for routes 8, 8A, 33, 45. All 25 Maleys, many recently overhauled and repainted, transferred en bloc from Dingle. Together with a group of Baby Grands and other Liners they displaced older cars and joined 269, 271, 881, 883, 916, 951 already at the depot together with snowplough 553.

Dingle: Having lost its allocation of Maleys, additional cars were needed for routes 21, 25 and the Dingle-Kirkby service. It is believed some early morning and late evening duties on the 21 may have been worked from Walton as well as some duties on the Dingle-Kirkby service. Photographs show many of Dingle's Standards were 'regulars' on the 21 during its final months.

Retained at Dingle	6, 309, 599, 864, 871, 877, 879, 884, 945, 952
Garston to Dingle	165, 313, 410, 638, 647, 734, 762
PAR to Dingle	358, 368, 462, 680, 765, 766, 819

Prince Alfred Road: Additional cars needed for all-day routes 46 and 49 and industrial services 38, 42 and 48.

Retained at PAR	40, 110, 130, 160, 162, 246, 260, 264, 274, 276, 296, 592, 710, 725, 745, 746, 748, 749, 751, 753, 754, 874, 875, 905
Dingle to PAR	916
Edge Lane to PAR	258, 265, 269-272, 275, 297
Garston to PAR	154, 167, 901, 925, 935, 937
PAR to Walton	20, 24, 646, 663, 684, 685

3 May 1950: Probable last day for post-match extras along Utting Avenue to Norris Green. These were lined up at the limit of the reserved track waiting for home-bound fans leaving Anfield. It is not known how many cars were involved, what fare was charged or whether transfers were available onto other routes. Also operation was confined to afternoon matches only and could be cancelled without notice. The public time-table stated "Patrons should note that these services can NOT be guaranteed when match-times clash with workers' peak hours."

6 May 1950: Seventy-three people injured when the driver of bus A102 tried to outpace 836 at one of the crossings on the central reservation near the Gillmoss pre-fab estate. Thousands witnessed this peak hour collision as they poured out of the factories opposite. 836 was swiftly towed away to Walton depot.

On the same day, the following cars were seen inside the Works.

Standards	6, 24, 62, 87, 141, 376, 451, 459, 638, 662, 684, 732, 753-755
Bogie cars	764, 771, 783, 791, 805, 806, 813, 815, 828, 836, 846, 856
Liners	157, 162, 168, 186, 188, 881, 885, 886, 893, 899, 918, 921, 935, 938, 945, 963, 970, 973, 976, 978, 982 (fire), 984, 986
Baby Grands	210, 204, 205, 208, 213, 216, 219, 221, 232, 248, 253, 254, 257, 268, 277, 278, 280, 286, 289, 298

The following cars were listed as in The Pool: 152, 878, 880, 900, 971, 972, 979

May 1950: To avoid renewing badly worn curves and junctions round the traffic island at Penny Lane, a new facing crossover was laid on Church Road for use by the 42, 48 and 49.

May 1950: With even basic maintenance as a low priority, breakdowns occurred all too frequently especially on cars based at Green Lane and Walton depots. At the Pier Head, a Marks Bogie has its trolley lowered as staff struggle to rectify the problem. Observing the situation is Jack Gahan (centre, holding mac) who recorded so much of the detailed information in this book and who also wrote a series of best-selling books covering different facets of Merseyside transport. Another local stalwart

was Walter Purdey who had a choice turn of phrase laced with frequent expletives. On one occasion, it is said Walter was on a Liner which broke down on Dale Street. Telling the driver he could repair the car, he began to dismantle the controller. After a few minutes, he stood back, looked at the driver and exclaimed ' the f… things bust' and walked off leaving the poor driver with a half-dismantled controller! No wonder Hatton Garden was wary of some of the enthusiasts. On the other hand, when Hall discovered that Ted Gahan, who was on the staff, was sending critical letters to the press, he was summoned to No 24. Ted was calm and respectful and reiterated the many concerns relating to the general state of the trams. Hall listened and said he would consider what he had said. *[Above] G W Rose/Online Transport Archive*

3 June 1950: During the early evening, 451 burst into flames on Muirhead Avenue. "A run-down Priestly Standard. Fault in the trolley leads caused fire – one of 36 during May/June – was this a record? The sight of burnt and charred wrecks being towed through the streets hardly helps the image". (JWG) The remains were scrapped in October. *Leo Quinn collection/Online Transport Archive*

June 1950: To make space for the major rehabilitation programme and clear the Works of cars awaiting repair, four unwired sidings were laid on a former flower bed between the west wall of the depot and the tracks flanking the east side of the Works. Using 608 as shunter, 211, 161, 981, 180, 781, 965, 155, 822, 832, 810 and 838 were relocated first. Known officially as 'the bank', for most enthusiasts it was 'the dump' especially when the sidings were later used to house vehicles awaiting scrap. *[Above] Martin Jenkins collection/Online Transport Archive*

When reconditioned each Liner and Baby Grand was rewired and, where needed, given new main pillars, panels, flooring, seats and sliding as opposed to wind-down windows (although this did not always happen). Some bodies also required internal bracing. To reduce the risk of body damage in accidents, most Liners were fitted with a more robust type of bumper. In most cases, glass rain guards, rubber gutters and ventilators were removed. The use of the side indicators was gradually phased out, most being removed internally and externally and, in some cases, painted or plated over sometimes with the screens still in situ. As an economy measure, the two-line via blinds on the Liners and Baby Grands were replaced by single line versions, although a handful of cars, including 178, still carried the two-line variety as late as 1953. By the end, it was claimed no two cars were alike! Anticipating future accident damage, additional pre-fabricated ends were built. This view was taken during an LRTL official visit on 18 June 1950. *[Below] W J Wyse/LRTA (London Area)/Online Transport Archive*

18 June 1950: Still campaigning vigorously against the closure programme, the LRTL held its annual tour on freshly-painted 881. Inspector Howard was again the driver and LRTL member T E Parkinson the conductor. Leaving Garston at 10.50am, the tram reached Roe Street via route 45 from where the tour officially started at 11.30am. A fast run to Kirkby via the 29 and 19 was followed by a visit to Fazakerley by way of the 44 and 22, after which there was lunch break in the city. Suitably refreshed, participants first explored every corner of the doomed Bootle routes before following the 36 to Walton and the 37 to Utting Avenue. From here it was back to the city via the 14 where there was a tea break. The final leg included a tour of the Works before a run along the 40 to Pagemoss and then via the 10C, 49 and 8 back to Garston where the tour ended at 7pm. The first view was taken at Kirkdale station and the second on Arkles Road, now used in both directions by cars on route 37. *[Opposite, top and centre] J H Price/National Tramway Museum; F N T Lloyd-Jones/Online Transport Archive*

92 | THE LEAVING OF LIVERPOOL

27 June 1950: Another air brake failure results in 861 being badly damaged when it was struck in the rear by 983 on Walton Hall Avenue during the morning peak. Albert Hibbert, the driver of the second tram, said 'How I managed to get out of the shattered cabin beats me. I stuck to the controls right to the end. The first tram was stationary and I was coming along to halt. I put on the air, but there was no response and I whipped the controls round, shutting off the electricity.' It is understood 861 was pushed onto the stub at the end of Utting Avenue so it could be examined prior to being towed to Edge Lane.

[Bottom left] Stan Watkins collection

27 June 1950: Peak turn out 405 cars, off-peak 210.

July 1950: When reporting the 881 tour in its columns, *Modern Tramway* referred to "the atrocious stretches of track in Bootle". Since expiry of the lease with Liverpool in 1942, Bootle Corporation had undertaken virtually no work on its 4¼ miles of track. Relations between the two Councils were fractious with Bootle threatening to give Ribble permission to replace the trams within its boundary if Liverpool did not pay sufficient compensation. Things became so bad that running times on routes 23, 24 and 28 had to be increased, schedules reduced and speed restrictions enforced. "Stanley Road appalling – dished joints, broken rail, missing check rail, raised setts and severe corrugation. Cars buck and sway violently." (JWG) The track in Litherland was slightly better as this was owned and maintained by Liverpool.

The schedule on the 23 and 24 was cut to every half hour and on the 28 to every 15 minutes. However, this had little effect on passenger numbers. For many years, local residents had used competing Ribble buses for a faster, more comfortable ride into the commercial heart of the city. As a result, outside peak times, many cars on Stanley Road were virtually empty. In partial compensation, the 17 was increased to every 10 minutes on weekdays, 15 minutes on Saturday and half-hourly on Sunday. This hardly helped as the 17 basically ran through industrial rather than residential neighbourhoods.

August 1950: Bellamy 558 and German trailer 429 moved from Garston to Edge Lane. 558 may have gone under its own power. Some reports suggest the move may have taken place in July.

14 August 1950: Despite the recent installation of new special work at the Whitechapel junction with Lord Street, route 21 used the worn track on Whitechapel and Paradise Street for the last time, with 958 making the final trip. The evening queue point on Whitechapel for extras to Dingle and Aigburth Vale also discontinued.

Track abandoned: St George's Place, Whitechapel, Paradise Street. Also the long disused track in Crosshall and Hood Streets as well as part of Roe Street was now officially abandoned.

15 August 1950: Route 21 rerouted between Byrom Street and Park Lane via Dale Street, Castle Street, South Castle Street and Canning Place. This led to a loss of some central area patronage. The evening extras from Whitechapel were replaced by additional departures from South Castle Street to Dingle or Aigburth Vale. Also evening extras run north from South Castle Street to Black Bull or Aintree.

Old Haymarket. Martin Jenkins collection/Online Transport Archive

Whitechapel. N N Forbes/National Tramway Museum

Paradise Street, at the junction with Lord Street, Church Street and Whitechapel ("Holy Corner") R G Hemsall/National Tramway Museum

2 September 1950: Routes 30, 34, 38 and 46 replaced. Of these, the 30 and 46 ran along Netherfield Road (North and South). "Netherfield Road track was not in good condition being well worn and laid in granite setts, which in many places stood proud of the sunken rails. The Maley & Taunton bogie frames of some streamliners seemed to have a very limited clearance and with the sinking tracks and the raised granite setts, drivers of these particular cars appeared to take extra care." (J A Bryant) During the blitz, this once tightly-knit area had lost much of its housing stock and, subsequently, more people had been rehoused so patronage had declined considerably.

The 30 ran every 24 minutes except in the evenings and on Sundays when it dropped to every half hour. Operated from Walton depot, it linked Pier Head to Aintree via Dale Street, William Brown Street, London Road, Moss Street, Netherfield Road, Walton Road, Rice Lane and Warbreck Road.

The main loss was the 46 Penny Lane to Walton (Spellow Lane). Cars in both directions used Smithdown Road, Upper Parliament Street, Grove Street and Crown Street as far as Moss Street. From here, northbound 46 went via Shaw Street, Netherfield Road and Walton Road but those heading south from Walton returned via Walton Road, St Domingo Road, Heyworth Street, Everton Road, Low Hill and Erskine Street. Frequency varied between 10 and 15 minutes. Operated from PAR, this important cross-city route, known to crews as "Spellow", was viewed as a tiring heavy-loader with many short-hoppers owing to the number of transfer points. It also served the University district, sports venues and places of entertainment. Despite the state of the track at the north end, 'Maleys' were regular performers.

Seaforth dock services 34 and 38 ran a handful of daily journeys which catered for those employed in and around the dock estate by providing direct links with various outer suburbs which would otherwise have involved at least one change of car or bus. With so few scheduled trips, these must have seemed ideal candidates for early conversion but the Corporation was in no rush to replace them as it derived 17% of overall income from the system-wide use of workmens' returns which were probably issued to virtually every passenger on the six Seaforth dock routes. Local councillors were also wary of any possible backlash against the loss of this valued concession. However, all post-war time-tables carried warnings that the services were 'liable to sudden alteration or cancellation'.

In the morning, some 34s showed 18 and in the evening 10C. In the morning, a minimum fare was in force on the 34 until the Edge Lane/Durning Road junction after which normal fares applied whilst in the evenings restrictions applied on both routes until reaching Commercial Road.

Owing to the City Engineer & Surveyor's deep-seated aversion to amending existing track layouts especially at junctions, morning workings on the 34 had, since closure of Sheil Road in June 1948, followed a circuitous route between Old Swan and Everton Valley by way of St Oswald's Street, Edge Lane, Paddington, Crown Street, Moss Street, Brunswick Road, Grant Gardens, Everton Road and St Domingo Road. On its last leg, the 34 followed the path of the 18A and 38 to Seaforth. The two weekday evening departures from Seaforth, together with the one Saturday mid-day journey, pursued a shorter, more direct course. On reaching Grant Gardens, eastbound 34s used Low Hill to access Kensington from where they made the long straight run to Longview Lane. Worked from Green Lane depot, it is not known how many workmen actually made this arduous round trip on a daily basis.

The 38 was basically a combination of the 46 and 18A linking Penny Lane and Seaforth via Smithdown Road, Upper Parliament Street, Grove Street, Moss Street, Brunswick Road (northbound), Low Hill, Erskine Street (southbound), Everton Road, St Domingo Road, Everton Valley, Kirkdale Road, Smith Street, Lambeth Road and Derby Road. Operated from PAR, there were just two early morning northbound departures Monday to Saturday balanced by one departure from Seaforth on Monday-Friday evenings and at mid-day on Saturdays.

ROUTE PROFILE
34, 38

LIVERPOOL CORPORATION PASSENGER TRANSPORT
P. 1282

INDUSTRIAL TRAM ROUTE
LONGVIEW **34** SEAFORTH

CONVERSION TO BUSES
Commencing Monday, Sept. 4th, 1950

Seaforth. Views showing either number are rare – so one with both is doubly rare! *N N Forbes/National Tramway Museum*

Rimrose Road. As so few cars had 34 on their screens, 681 is showing 10C, Longview Lane. *N N Forbes/National Tramway Museum*

Penny Lane. With just a card in the window, this Standard prepares to leave for the cross-city run on route 38 to Seaforth. *J S Webb/National Tramway*

Grant Gardens. A grimy 'Maley' works an evening journey towards Seaforth. *E A Gahan*

Journey time	Approx 45-50 minutes (34), 34 minutes (38)
Last day of operation	2 September 1950
Last cars	Not known

1950 | 95

ROUTE PROFILE
30, 46

Penny Lane. The cross-city 46 started its northbound journey from Penny Lane. *Leo Quinn/Online Transport Archive*

Pier Head. The 30 started from the North Loop at Pier Head. *Peter Mitchell*

William Brown Street, with the Wellington Column in the background. *E A Gahan/Online Transport Archive*

The steep single track and loops section along narrow Netherfield Road fascinated Norman Forbes. These are just a few of his remarkable views. Netherfield Road/Brow Side. The 46 is working north and the 25 south (see text). *N N Forbes/National Tramway Museum*

Netherfield Road/St George's Hill. This section was used by peak hour cars from Walton depot, working into the city, to take up service on other routes, such as 14A. *N N Forbes/National Tramway Museum*

LIVERPOOL CORPORATION PASSENGER TRANSPORT P. 1274
TRAM **30 46** ROUTES
38
CONVERSION TO BUSES
Commencing:—
Routes 30, 46—Sunday, Sept. 3rd 1950
Route 38 Monday, Sept. 4th 1950

96 | THE LEAVING OF LIVERPOOL

Netherfield Road. A southbound 25 leaves the loop by City Hospital North followed by a car proceeding to the city to cover a peak hour duty. With the closure of this section, southbound 25s were rerouted via St Domingo Road. *N N Forbes/National Tramway Museum*

Netherfield Road/Thomaston Street. *N N Forbes/National Tramway Museum*

Netherfield Road/Everton Valley *N N Forbes/National Tramway Museum*

Carisbrooke Road, Walton. *Martin Jenkins collection/Online Transport Archive*

Journey time	32 minutes (30), 30 minutes (46)
Last day of operation	2 September 1950
Last cars	Not known
Track abandoned	Netherfield Road (North and South)

1950 | 97

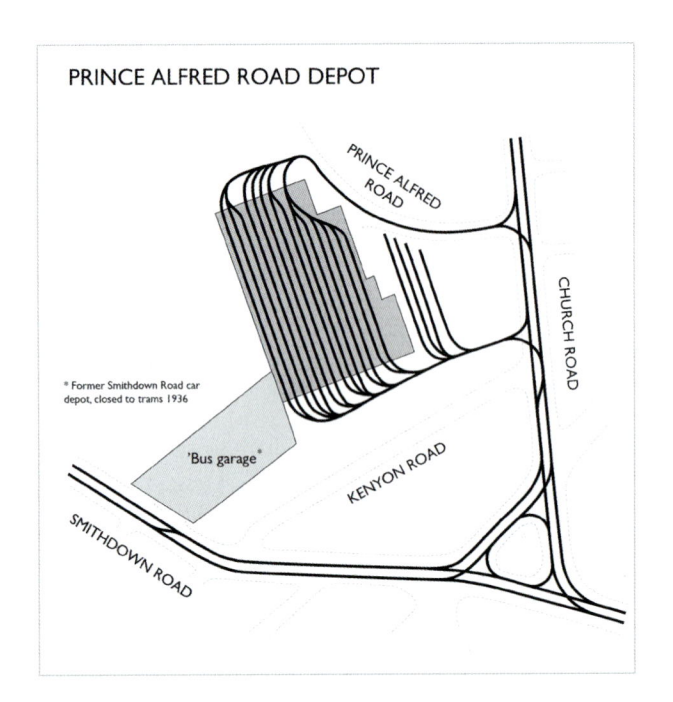

PRINCE ALFRED ROAD DEPOT

Former Smithdown Road car depot, closed to trams 1936

'Bus garage'

Prince Alfred Road Depot: This was the first depot to lose its trams. With replacement of the 38 and 46, routes 42, 48 and 49 were transferred to Edge Lane. In the post-war years, PAR staff took steps to prevent defective cars going to Edge Lane preferring to undertake repairs themselves. They also tackled the problem of drooping platforms on the Liners and Baby Grands by bending their steel platform bearers back into alignment. Skilled technicians kept the Liners, especially the Maleys, on the road and some may well have followed the Maleys to Dingle and then to Garston. Following its official closure, it is possible PAR stored cars for a few more months as an official document dated 6 February 1951 states "PAR cleared of all trams". Former driver, Bill Peters, recalls in his book *Busman*, that old trolley ropes were reused as chains in the depot lavatories! PAR became a bus depot and served as such until October 1986.

The first view shows recently refurbished and repainted English Electric bogie car 767 in the depot yard, whilst in the second view, 359 is being manoeuvred at the junction of Church Road and Prince Alfred Road by shed men.

[Below, left] Stan Watkins; N N Forbes/National Tramway Museum

3 September 1950: Southbound 25s now rerouted via Kirkdale Vale, St Domingo Road, Grant Gardens and Erskine Street in lieu of Walton Road, Netherfield Road and Shaw Street. Route 31 reduced to operating between Walton (Spellow Lane) and the Pier Head and rerouted via William Brown Street and Dale Street to replace the 30.

September 1950: Last cars in old red livery withdrawn. Walton-based 137 and 577 were regulars especially at rush hours. Near the end, 137 passes the boarded up west side of Dingle depot on a cross-city 21 to Aintree. It is possible it may have been on loan to Dingle, which operated the 21, or, as suggested earlier, some duties on the 21 may have been operated from Walton. Now the oldest surviving passenger car, 577 of 1913 is at Seaforth in August 1950 probably preparing to leave on one of the dockers' services 35, 36 or 37. *[Above] Alan B Cross; J S Webb/National Tramway Museum*

98 | THE LEAVING OF LIVERPOOL

20 October 1950: LRTL member T E Parkinson recorded the following Standards still available for service out of a total fleet of 465 cars.

Handbrake	1, 6, 11, 14, 15, 18, 22, 24, 26, 29, 30, 36, 40, 49, 50, 62, 65, 67, 69, 76, 84, 85, 90, 92, 99, 104, 109, 110, 118, 130, 131, 133, 135, 140, 144, 302, 309, 310, 313, 321, 324-327, 344, 359, 368, 373, 376, 380, 389, 400, 410, 441, 450, 462, 592, 599, 637, 645-648, 650, 652, 659, 663, 664, 667, 674-682, 684, 685, 687, 688, 693, 695-697, 700, 701, 703, 710, 712, 716, 718, 719, 734, 739
Airbrakes	5, 12, 35, 38, 81, 87-89, 91, 95, 97, 98, 101, 114, 128, 147, 328, 338, 342, 343, 353, 440, 459, 673, 722-727, 729, 731, 732, 745-755,

28 October 1950: Dense 'pea-souper' fog envelopes Merseyside. In the choking gloom, 275 collided with 261 on the single track on St Oswald's Street, 368 and 848 met on a section of single line head-on in Townsend Lane, 909 hit bus A267 at the Carlton junction and 176 was destroyed by fire whilst working a 49 near Olive Mount hospital at 6.20am. One of its motors ignited and the heat was so intense that flames leapt 20 feet into the air and distorted the overhead. It took over two hours to move the ruined shell to Edge Lane, its lightweight trucks being transferred subsequently to 961.

18 November 1950: As part of a new traffic scheme aimed at alleviating congestion on the approaches to the Mersey Tunnel, the track in Victoria Street had to be abandoned.

The last car to leave was 722 working a football special bound for the siding on Walton Breck Road at 2pm. The remarkable short length of five parallel tracks at Old Haymarket (seen in the background) was now lost. Note the stylish Art Deco street lamps, including the one illuminating the Tunnel entrance, and the disused track along Whitechapel. It is not known when Walton Breck Road siding was last used as some football extras may have continued running to and from the Pier Head until abandonment of the 37 in February 1951. *[Above] N N Forbes/National Tramway Museum*

As part of the revised traffic arrangements, Old Haymarket now became a stub terminal, its three tracks being accessed from William Brown Street. Most of St John's Lane was also abandoned.

From an unknown date (probably 20 November), Old Haymarket became an evening and Saturday mid-day peak hour boarding point for routes 6, 6A, 13 and 13A. Prior to this, it had been used by extras showing 10A, 10B, 10C, 19, 19A, 29A, 30 and 31 although none of these duties appeared in any published time-table. For example, on 13 July 1950, 668 is working a 31 to Walton and 755 a 10C to Longview Lane. Sometimes up to three inspectors, operating from a small cabin complete with phone contact, were needed to marshal the arriving cars and gathering queues. *[Overleaf, top] Peter Mitchell*

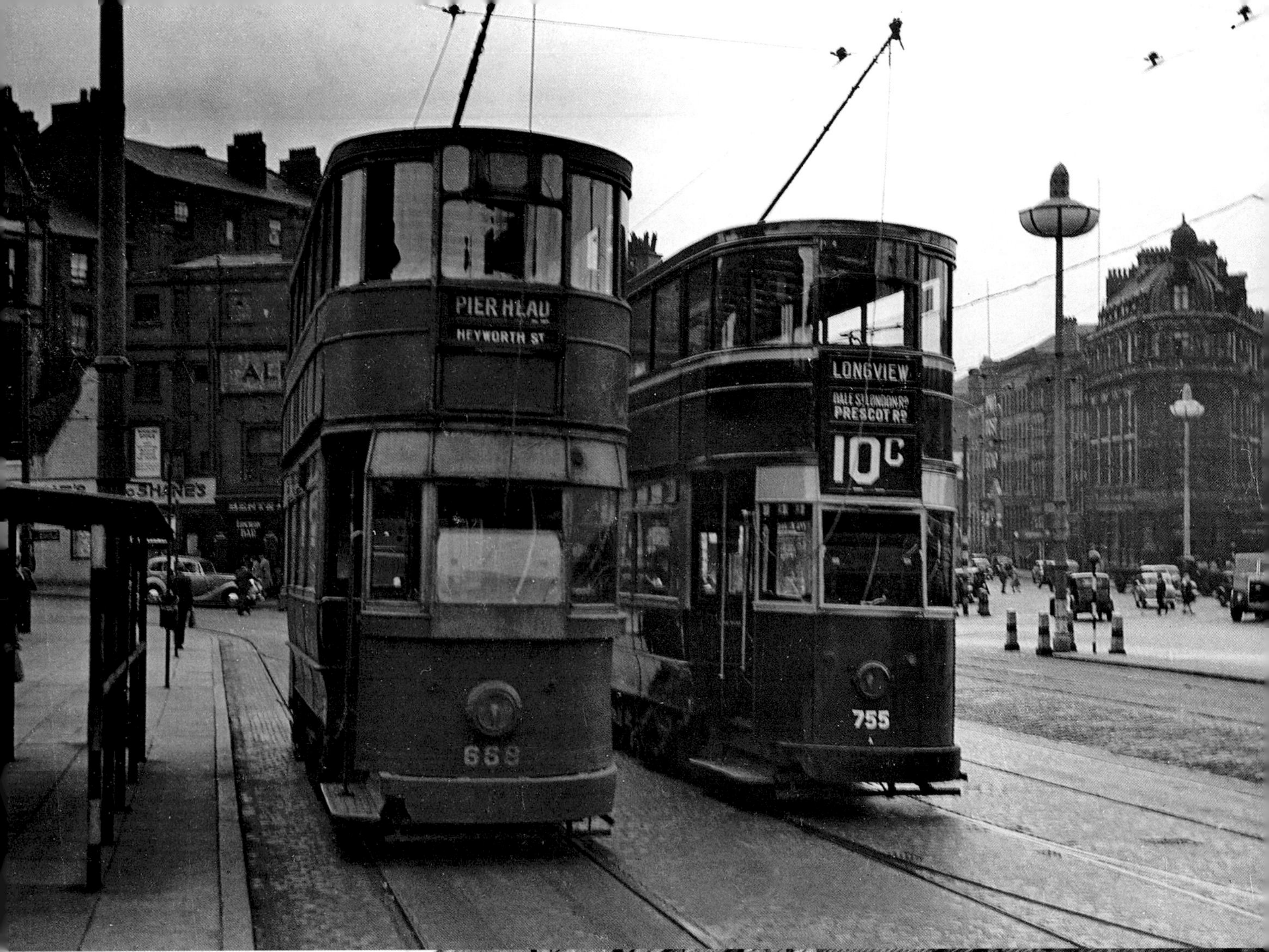
PIER HEAD
HEYWORTH ST
LONGVIEW
PRESCOT Rd
10C
755

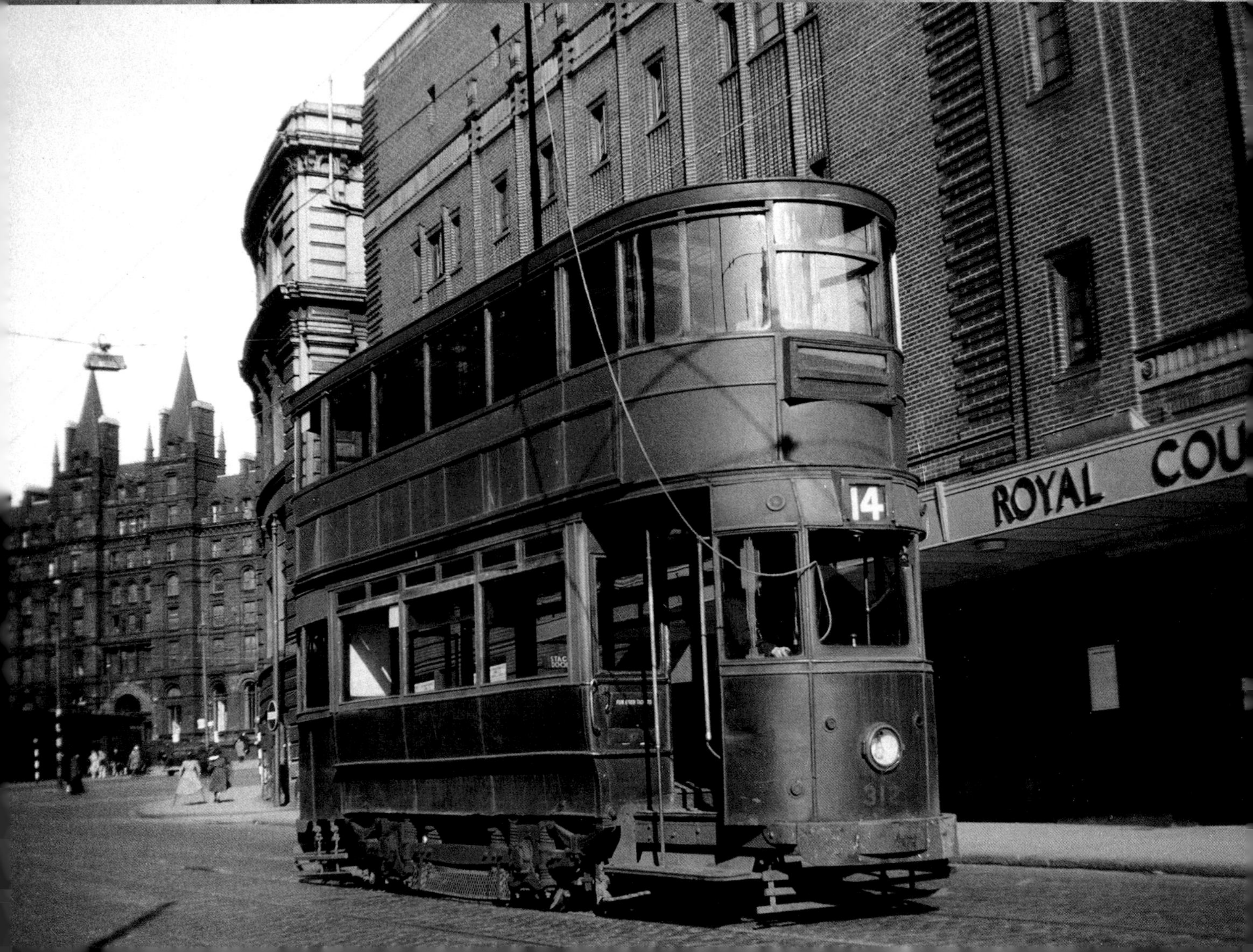
14
ROYAL

On another unknown date, routes 6/6A and 14/14A swapped their evening boarding points with the 6s moving to Roe Street and the 14s to Old Haymarket. Shortly before the switch, 312 of 1922 waits to reverse at Roe Street. This was one of a small batch of Standards which had route number boxes built into the driver's windscreen. Others included 85, 120, 321, 359 and 638. 312 was withdrawn soon afterwards. *[Left, bottom] N N Forbes/ National Tramway Museum*

November 1950: Following cars assigned to Litherland depot shortly before it closed to trams: 5, 12, 25, 31, 47, 48, 51, 53, 57, 64, 65, 68, 73, 75-77, 81, 91, 94, 101, 108, 114, 116-118, 120, 124, 203, 278, 342, 450, 676, 817, 833, 846 and 851. Some of these had been out of service for a considerable time, for example war-damaged 31 had been gathering dust for some nine years. During its conversion into a bus garage, some duties may have been covered by Walton-based cars. Latterly, owing to the atrocious state of the track, only four-wheel cars tended to be assigned to the routes using Stanley Road

2 December 1950: Routes 16, 23, 24 and 28 replaced and peak hour only route 22A discontinued. Trams finally stop using the life-expired track in much of Bootle. All but the 22A were operated from Litherland.

Although carrying residential traffic especially in Litherland and along the Stanley Road/Scotland Road corridors, these routes were mainly used by those employed in the Dock Estate and the myriad of nearby warehouses, factories, railway depots, coal and timber yards where there was still a great deal of war-damaged property.

On leaving the stub terminal in Wellington Road, Litherland, the 16 and 28 shared a section of single track and loops along Bridge Road and Linacre Road. They then ran together along Stanley Road as far as the junction with Commercial Road where the 16 reached the Pier Head by way of Vauxhall Road, Tithebarn Street and Chapel Street whilst the 28 arrived at South Castle Street by way of Stanley Road, Scotland Road, Byrom Street, Dale Street and Castle Street. There were no known short workings except for depot journeys. The 16 operated every 20 minutes throughout the week.

The 23 and 24 linked Seaforth with South Castle Street, the 23 reaching Stanley Road via Rimrose Road and the single track and loops section on Strand Road and the 24 via Knowsley Road. From Stanley Road, they followed the same path as the 28. If the cramped terminal stub on Regent Road, Seaforth became over-crowded, a duty inspector could turn a car on a crossover in nearby Rimrose Road.

Passengers travelling in either direction between Litherland depot and Seaforth were charged a flat fare of 1½d although a 2d workman's return was available before 8am. All these journeys involved a reversal on Stanley Road.

Litherland depot: Three cars, each with a long association with the depot, are seen in section 'A' in October 1950 by which time sections 'B' to 'D' were being converted for use by buses. Having been extended and rebuilt in 1939, the depot is believed to have had 16 short length tracks. No Liners ever seem to have been allocated and any snowploughs were usually housed on the northern most track in section A. To date, it has not been possible to establish the exact date when it last housed trams. A photo taken in late November shows the wiring into section 'A' had been cut suggesting operation of the 16/28 and 23/24 had

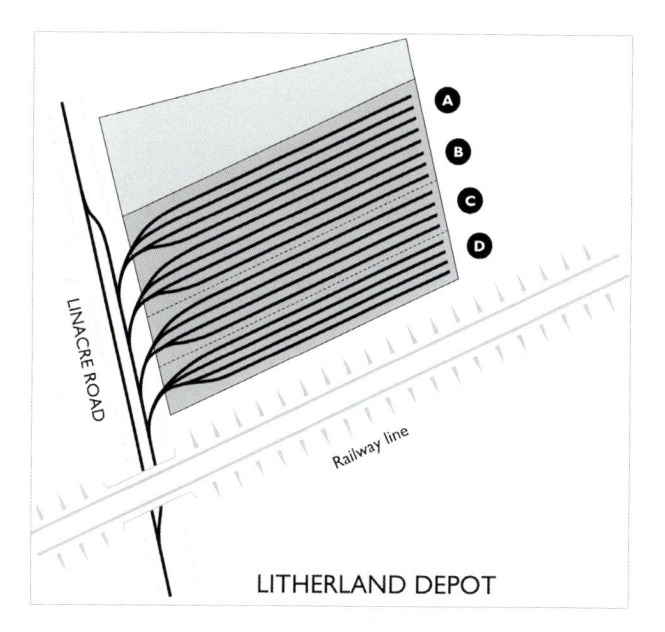

LITHERLAND DEPOT

been transferred to Walton. Furthermore, on the final Saturday, Norman Forbes took several scenes featuring Walton-based cars such as 96, 280 and 286 most running with blank indicators. At some unknown date, operation of the 17, 18 and 18A was also transferred to Walton and, like the 16/28 and 23/24, cars were probably still crewed by Litherland men who had to go to Walton to sign on. As a bus depot, Litherland closed entirely in 1986.
[Above] G F Douglas, courtesy A D Packer

3 December 1950: "Snow showers commenced about 12.30am resulting in Sunday service being worked by old Brill cars, Baby Grands, Cabins and Marks. Very few streamliners left running. Policy appeared to be to keep them off the road so they would be in good enough condition to work Monday rush hour service. First time old Brills had worked on Sundays for some time." (JWG)

ROUTE PROFILE
16, 28

Wellington Road, Litherland. *A D Packer collection*

Bridge Road, Litherland. *N N Forbes/National Tramway Museum*

Linacre Road, passing Litherland depot. *N N Forbes/National Tramway Museum*

102 | THE LEAVING OF LIVERPOOL

Stanley Road/Knowsley Road. *N N Forbes/National Tramway Museum*

Stanley Road/Strand Road. *N N Forbes/National Tramway Museum*

Stanley Road, Bank Hall. *N N Forbes/National Tramway Museum*

Vauxhall Road at the Hopwood Street stop. *A W V Mace/National Tramway Museum*

Vauxhall Road, with Liverpool's "Bridge of Sighs" dominating the background (part of the Tate & Lyle complex). *N N Forbes/National Tramway Museum*

ROUTE PROFILE
23, 24

Knowsley Road, Seaforth, passing prefabs. *N N Forbes/National Tramway Museum*

Knowsley Road, Seaforth, passing under the Liverpool-Southport electric railway. *N N Forbes/National Tramway Museum*

Rimrose Road, Seaforth. Note the Bootle Corporation tower wagon on the right. *N N Forbes/National Tramway Museum*

104 | THE LEAVING OF LIVERPOOL

Strand Road. *N N Forbes/National Tramway Museum*

Strand Road, again passing under the Liverpool-Southport railway.
N N Forbes/National Tramway Museum

Scotland Road/Stanley Road, known as "The Rotunda" after a famous theatre that was bombed. During the period that Litherland was being converted to a bus garage, some or all of its duties were transferred to Walton. This Walton-based car does not have the correct indicators and as a result has a card '24' in the window. *N N Forbes/National Tramway Museum*

Derby Square. *E A Gahan/Online Transport Archive*

Journey time	36 minutes (16), 38 minutes (28) journey times 34 minutes (23, 24)
Last day of operation	2 December 1950
Cars seen in service on the last day	5, 114, 128, 342, 864
Last cars	864 (16), not known (others)
Track abandoned	Wellington Road, Bridge Road, Linacre Road, Stanley Road, Regent Road, Knowsley Road, Strand Road, part of Commercial Road, Vauxhall Road

LIVERPOOL CORPORATION PASSENGER TRANSPORT

TRAM ROUTES

LITHERLAND **16, 28** CITY

SEAFORTH **23, 24** CITY

CONVERSION TO BUSES

Commencing – Sunday, December 3rd

ROUTE PROFILE
22A

The 22A was another little photographed industrial route. Operated from Walton, it provided a direct link between Pier Head and Fazakerley via Chapel Street, Tithebarn Street, Vauxhall Road, Commercial Road, Melrose Road, Hale Road, Rice Lane and Longmoor Lane. Latterly, there were five departures from the city on Monday-Friday mornings plus one at 8.15pm. In the reverse direction, there was a single early morning journey from Fazakerley Monday to Saturday plus two from Black Bull only on Monday to Friday at 1.33pm and 4.33pm. No wonder there are few photographs!

Journey time	40 minutes (approx)
Last day of operation	2 December 1950
Last cars	Not known
Track abandoned	None

Fazakerley. Martin Jenkins collection/Online Transport Archive

Hale Road, car 333. N N Forbes/National Tramway Museum
Jack Gahan rode on this car and wrote a vivid report (see below).

"Route 25 Aigburth Vale to Walton, car 333, Saturday, May 20th, 1950. This car, a Priestly Standard (4-wheel Brill 7ft 6in truck) of 1922 is a complete ruin. The body is warped and sway-backed, and all joints are loose. The longitudinal wooden seats are splitting away from the side, and when someone sits down at one end, the opposite end comes up! The car appears to be falling inward, and you duck sideways to get through the door at one end. The window ledges on each lower deck window are arch-shaped, as the car travels (quite fast) the body creaks and leans from side to side, while the platforms bounce up and down. A piece has had to be cut in the wartime windscreen so that the brake handle can make a full turn. The road can be seen through the floorboard edges. Comments from passengers were not complimentary. Unfortunately, many of the adjectives used to describe old trams, such as 'grinding', 'rocking', 'swaying', 'pitching', 'tossing', 'creaking', 'jolting', 'roaring' and 'crashing' could all-too-easily be applied to the Liverpool cars of this period." **(JWG)**

6 December 1950: 780 was one of convoy of four fully-laden trams leaving the Kirkby 'Admin' loop at 5.30pm when it 'jumped the track after crossing the points and crashed between two standards supporting the overhead wire'. As a result, it toppled over and totally blocked South Boundary Road which remained closed until the wreckage was cleared by 9pm. 43 people injured, 11 seriously but driver William King and conductor, N de Silva escaped unscathed. Another tram commandeered as temporary casualty station. This view shows how the top deck was cut away to access the injured. *[Above] Martin Jenkins collection/Online Transport Archive*

106 | THE LEAVING OF LIVERPOOL

30 December 1950: Replacement of route 17 Pier Head to Seaforth via Chapel Street, Tithebarn Street, Pall Mall, Great Howard Street, Derby Road and Rimrose Road. Although this all-street track route ran fairly close to the Dock Road most people preferred the faster Liverpool Overhead Railway to access the bustling area round the north end docks. Much of the final day took place during periodic blizzards. No known short workings.

On the same date, route 22 ceased to use Great Crosshall Street, Tithebarn Street and Chapel Street thereby eliminating a single track section in the heart of the business district. Trams no longer served Exchange Station. Here it's all the 2s! – Standard 22 of 1922 on route 22. The car is using the complex Tithebarn Street/Great Crosshall Street junction. Although points were laid into Marybone when the junction was last renewed, they were never used. *[Above] J S Webb/National Tramway Museum*

31 December 1950: 22 rerouted via Byrom Street and Dale Street.

Cars believed to have been rebuilt or rehabilitated during 1950: 153, 157, 158, 168, 174, 208, 213, 221, 232, 246, 253, 254, 258, 261, 272, 275, 277, 279, 883, 884, 887, 893.

Cars believed to have been withdrawn and sold to Maden & McKee during 1950: As in previous years, the precise date on which cars last ran in service is often not known nor the exact date on which they were broken up and burnt. After withdrawal, some were laid up for quite some time possibly as a source for spares.

Month	Cars
January	102 (collision), 301, 572 (scrapped March 1950), 651, 654 and 709 (scrapped March 1950)
February	Snowplough 513
March	25, 73, 94, 116, 134 (snowplough), 331, 383, 507, 527, 539, 540, 546, 565 (snowploughs), 585, 605 (decorated ceiling, cushioned longitudinal seats, narrow frame truck), 672, 694, 707 and 741 (collision)
April	57 (survived machine-gun attack during the Blitz), 105, 112, 454 and 744.
May	88 (collision), 469, 608 (used latterly as Edge Lane shunter)
June	638
July	339 (collision)
August	20 (collision, scrapped October 1950), 28, 50, 63 (scrapped October 1950), 75, 82, 129, 141, 305, 307 (scrapped November 1950), 415, 660, 662 and 669 (scrapped October 1950)
September	137 (scrapped October 1950), 312, 577 (scrapped October 1950) and 730
October	34, 333, 358, 378 (scrapped 1951), 602, 658 and 668
November	135, 138 and 731

Also withdrawn during the year but month unknown: 62, 77, 130, 341, 400, 605, 690, 698, 714 and 737.

ROUTE PROFILE
17

Derby Road. *N N Forbes/National Tramway Museum*

Derby Road. *N N Forbes/National Tramway Museum*

Pall Mall/Tithebarn Street. *N N Forbes/National Tramway Museum*

108 | THE LEAVING OF LIVERPOOL

Great Howard Street/Boundary Street. *N N Forbes/National Tramway Museum*

Chadwick Street, passing under the railway lines just outside Liverpool Exchange station. *N N Forbes/National Tramway Museum*

LIVERPOOL CORPORATION PASSENGER TRANSPORT

TRAM ROUTE

SEAFORTH 17 PIER HEAD

CONVERSION TO BUSES

Commencing Sunday, Dec. 31st, 1950

Journey time	34 minutes
Last day of operation	30 December 1950
Cars seen in service on the last day	38, 168, 343, 778, 858. The appearance of 168 indicates that Walton depot had probably operated the 17 during its final month as there is no record of Liners being allocated to Litherland during the post-war period. 38 and 778 were also based at Walton.
Last car	Believed to be 778
Track abandoned	Part of Georges Dock Gates, Chapel Street, Tithebarn Street, Great Howard Street and part of Great Crosshall Street. At the same time, the long-disused track in Hatton Garden and Moorfields was officially abandoned.

1950 | 109

ROUTE PROFILE
17 *(continued)*

Top: Tithebarn Street. *Michael H Waller/Online Transport Archive*

Above, left: Chapel Street/Rumford Place. *N N Forbes/National Tramway Museum*

Above, right: Chapel Street/Lancelot's Hey. *N N Forbes/National Tramway Museum*

Left: Pier Head, North Loop. Nicknamed 'The Crystal Palace', 742 was the first totally enclosed car in the fleet. *N N Forbes/National Tramway Museum*

CHAPTER 8 | 1951

Such was the determination to dispose of the trams, Stage One was completed ahead of schedule with most older trams and worn out track being eliminated. Cllr Armour stood down as Chairman of the Transport Committee. Initially he had supported development of the system but, faced by post-war austerity, he followed the national trend and supported the conversion programme. However, when it was clear the replacing buses were failing to meet expectations, he was not prepared to consider retaining any part of the tram system which was described in *Modern Tramway* as "having been brought up to a much higher standard of maintenance than was common a few years ago".

In the middle of the year, Alderman Sheenan, leader of the City Council, bemoaned the rise in fuel tax estimating it was costing an extra £216,000 a year. He lambasted the increases as 'monstrous'. In an attempt to raise nearly £500,000, fares were raised in March and again in October. For too long, fare box revenue had failed to cover the spiralling cost of wages, fuel, labour and materials, a situation exacerbated by staff shortages, increasing militancy and absenteeism, all of which coincided with a marked drop in ridership. The trams had been profitable until 1946 but ever since it was claimed they were losing £500,000 a year. Overtime was also a problem as an additional 500 platform staff were needed to supplement the nearly 4000 men and women employed as bus and tram drivers and conductors.

Track relaid during the year: all or part of Ranelagh Street, Muirhead Avenue, Broadgreen Road, Towerlands Street, the Edge Lane/Mill Lane junction and the Edge Lane/St Oswald's Street junction (see later).

1 January 1951: Great Crosshall Street continued as peak hour terminus for routes 2, 22, 44 and 44A.

2 January 1951: Following the recent closures, more Baby Grands were again assigned to Garston: 203, 207, 242, 247, 258, 261, 269, 271, 273, 283 and 287, where they joined 440, 868-870, 872, 873, 878, 879, 881, 883, 884, 901, 903, 905, 916, 918-942, 947, 951, 955, 962, 966, 968, 974, 976, 979 and 992. The Baby Grands were intended for the 8A and 45 but again they proved troublesome.

29 January 1951: To allow for reconstruction of the North Loop at the Pier Head to provide loading spaces for the growing number of buses, trams were removed and the following routes relocated. Officially the 22 was moved to the Centre Loop and the 2, 13/13A, 14/14A, 19/19A, 31 and 44/44A to the South loop. However, photographs exist of 22s on the South Loop. At the same time, the 39 and 40 moved from the Centre to the South Loop and the boarding point for the 8 and 8A to the section of track linking the South and Centre loops. Finally, the 33 and 45 loaded on the post-war siding on the approach to South Loop. The image above shows four cars on the North Loop shortly before abandonment. *Martin Jenkins collection/Online Transport Archive*

Track abandoned: Part of George's Dock Gates and St Nicholas Place

Views of trams on the short length of George's Dock Gates are rare. In the latter part of the war, scores of US fighter planes arrived at the north docks. These often progressed through the streets under police escort which resulted in major delays. The Liverpool Overhead Railway dominates this view of one such hold-up at the junction of Water Street and George's Dock Gates. *[Overleaf, top] Brian Martin collection/MTPS*

30 January 1951: Routes 13/13A, 14/14A, 19/19A and 44/44A rerouted via New Water Street.

The reorganisation proved premature with lines of cars on New Water Street trying to access the remaining loops especially at peak times when up to four inspectors were required. Occasionally, drivers made an additional circuit of a loop to meet their correct departure time. *[Overleaf, bottom] Alan B Cross*

112 | THE LEAVING OF LIVERPOOL

Sometimes, cars on the 8/8A could be 'trapped' by those occupying the Centre loop. In the reorganisation, the 29s were also included in the routes to be transferred to the South loop but this never happened. To ease pressure on the over-loaded South Loop, the 2 and 31 soon moved to the Centre loop.

[Above] J W Gahan

3 February 1951: Further reshuffle at Garston in an attempt to reduce the number of Baby Grands. Allocation recorded as Standard 440 plus 55 Liners and six Baby Grands.

> "One of my favourite outings was to take the Overhead to Seaforth and walk down to watch all the trams coming and going at the nearby terminus. There always seemed to be so much happening – conductors swinging trolleypoles, crews going into the small 'office' and emerging with cans of tea and inspectors shouting instructions. Every few minutes there was a tram arriving or departing. I only wish I had worked out some of the industrial services as I usually visited on a Saturday when they would certainly have appeared around mid-day. I remember going back into town on a 16 and later a 17. On my last visit sometime during the 1950 October half-term, I boarded an 18. I recall the conductor coming upstairs and dropping down the trap door that prevented access to the stairs at the driver's end. What a ride – couple of long straight runs then all manner of twists and turns as the car made its way through Walton and Everton. I remember the driver working the handbrake and the car lurching and dipping through a range of worn junctions. Where would I end up? When we got to Breckfield Road, I recognised where I was and caught a 13 or 14 back into town. Interestingly, my mother had used these routes when she taught at Anfield Road school in the 1920s. However, she was not very complimentary about the old Bellamys saying the pitching and tossing often made her feel sick!" **(MJ)**

10 February 1951: The 18 was the last all-day route in Bootle. From Seaforth, this outer-suburban service followed a twisting all-street track course via Rimrose Road, Derby Road, Sandhills Lane, Lambeth Road, Smith Street, Whittle Street, Kirkdale Road, Everton Valley, Walton Breck Road and Robson Street to terminate at Breckfield Road North (shown as Breck Road, Breckfield Road or Breckfield Rd. on screens). Here passengers were able to transfer onto the 13/14 group of routes. Other than depot journeys, the only known short-working was to the crossover on Derby Road catering for the workforce at the Grayson, Rollo and Clover dockyard. The basic headway was every 15 minutes on weekdays decreasing to 20 or 30 minutes on Sundays.

Industrial service 18A also operated for the last time. Running entirely on street track, it shared the same path as the 18 as far as Everton Valley where it turned south along St Domingo Road, Heyworth Street and Everton Road to terminate at Grant Gardens (Everton Road or Grant Gardens/Everton Road on screens). Here, passengers could transfer onto the West Derby Road routes. Latterly, cars left Grant Gardens every 10 minutes on Monday to Friday mornings and from Seaforth every 10 minutes in the evenings. On Saturdays, morning departures were the same but with only six mid-day departures in the reverse direction. There were two scheduled Sunday workings.

The lengthy outer-suburban industrial service 36 linking Seaforth to Lower Lane or Gillmoss duplicated the 18/18A to Sandhills Lane after which it diverged along Commercial Road, Melrose Road and Hale Road in order to access County Road, Walton Road and Kirkdale Vale where it turned sharply onto Everton Valley to proceed via Walton Lane and Walton Hall Avenue. Latterly, there were four Monday to Saturday early morning departures from Gillmoss and two from Lower Lane matched by a single working from Seaforth, Monday to Friday at 5.35pm and 12.05pm on Saturdays.

Industrial service 37 linking Seaforth to Utting Avenue East followed the 18/18A to Everton Valley from where it headed north-east along Walton Breck Road, Arkles Road, Utting Avenue and Utting Avenue East. Latterly, the time-table showed six inbound departures Monday to Saturday mornings whilst the evening departures replicated those for the 36. For some two years, cars on the 37 picked their way at 15mph along the increasingly decrepit single line on Utting Avenue.

To protect those using workmen's return tickets, minimum fares were in force on the 35, 36 and 37 as far as Commercial Road in the evening and as far as Hale Road (36) and Broad Lane (37) in the morning. There were no known short-workings except for depot runs.

Latterly, all four routes worked from Walton depot. However, there is no record of the route they took in the morning to the outer termini of the 36 and 37. One enthusiast recalls that 37s arriving at Utting Avenue East in the evening turned onto Dwerryhouse Lane for the run back to Walton depot by way of Walton Hall Avenue or possibly to take up another duty. Latterly, the numbers 35 and 36 were rarely shown as they had not been added to the screens on the surviving trams so route cards were sometimes displayed. It is assumed 'regulars' would have known departure times, routing and ultimate destinations.

15 February 1951: Last Bootle trams ran on this cold, damp Thursday evening. The final route was the 35 linking Seaforth to Fazakerley. Having duplicated the 36 as far as the top of Hale Road, 35s then turned north along Rice Lane and Longmoor Lane. This route had even fewer scheduled workings than the 36 and 37 – just a couple of early Monday to Saturday trips

ROUTE PROFILE
18, 18A, 36, 37

Breckfield Road. *A D Packer*

Everton Valley. *Stan Watkins*

Kirkdale Road/Whittle Street *N N Forbes/National Tramway Museum*

Turning into Whittle Street from Smith Street. *N N Forbes/National Tramway Museum*

Lambeth Road/Stanley Road. *N N Forbes/National Tramway Museum*

Derby Road/Sandhills Lane. *A S Clayton/Online Transport Archive*

Derby Road police station. *N N Forbes/National Tramway Museum*

Regent Road, Seaforth. *A D Packer*

114 | THE LEAVING OF LIVERPOOL

Melrose Road. This is the only known view of a car showing 36.
N N Forbes/National Tramway Museum

Views of cars showing 37 are also relatively rare. 841 is at the limit of the stub on Regent Road, Seaforth on 15 May 1949 just before setting off on an evening run to Utting Avenue East. *E A Gahan/Online Transport Archive*

Utting Avenue. Note the 15mph speed restriction imposed by the MOT.
[Left] E A Gahan

Journey time	30 minutes (18/18A)
Last day of operation	10 February 1951
Cars seen in service on last day	18/18A: 104, 770, 775, 778, 815, 867, 963 36: 14 37: 682
Last cars	963 (18); 867 (18A, the 1.30pm departure from Seaforth); 14 (36, the 12.05pm from Seaforth); 682 (37, the 12.05pm from Seaforth)
Track abandoned	Lambeth Road, Smith Street, Whittle Street. The surviving track on Utting Avenue, Arkles Lane and Walton Breck Road was probably abandoned as from Sunday 11 February although one local enthusiast recalls that for 'several months' after the conversion of the 37, some drivers still may still have taken this route when returning to the depot from Utting Avenue East or when terminating at Broad Lane with a short working 13A or 14A so the final closure date cannot be confirmed. 817 was recorded short-working to Broad Lane and reversing in Utting Avenue on Friday 9 February. Probably the last car to use this crossover.

Utting Avenue/Broad Lane. In the first view, 678 and 667 have completed the painful progress from Walton depot along the remaining single line on Utting Avenue. In the second view, 677 turns towards the city on a peak hour 14A, whilst 678 has used the former 43 crossover, but in the reverse direction, in order to proceed towards Utting Avenue East. Short workings on the Breck Road services here showed 'Broad Lane' and not 'Utting Avenue'. *[Above left and left] N N Forbes/National Tramway Museum (both)*

1951 | 115

ROUTE PROFILE
35

Hale Road, approaching junction with County Road. One of these cars will be on the 35 and the other the 36. Neither carry any form of route or destination display. *N N Forbes/National Tramway Museum*

Commercial Road. *N N Forbes/National Tramway Museum*

Journey time	Estimated 35 minutes
Last day of operation	15 February 1951
Cars seen in service on last day	834
Last car	834 left Seaforth at 5.35pm
Track abandoned	Regent Road, Rimrose Road, Derby Road, Commercial Road, Melrose Road, Hale Road

116 | THE LEAVING OF LIVERPOOL

from Fazakerley one of which also ran on Sundays whilst, in the reverse direction, there was a single Monday to Saturday departure from Seaforth at exactly the same time as the 36 and 37. Again there are no known short-workings except for depot runs and there is no known photograph of a car with 35 on its screen. Until the number was introduced in January 1947 cars probably showed 22 when heading for Fazakerley and 36 in the other direction.

Liner 990 selected as the official last car. On board were the Lord Mayor of Liverpool, the Mayor of Bootle, civic representatives and members of both councils including Alderman Bidston, one of those responsible for promoting the conversion programme. In 1948, it was claimed it would have cost nearly £400,000 to relay the track in Bootle. When 990 left Seaforth at 6.30pm, the local press reported that "cheering crowds lined the route and passing buses flashed headlights". After travelling to the Bootle boundary at Dacre Street, everyone alighted and transferred onto two buses, L459 and L466, for a reception at Bootle Town Hall to which the driver and conductor of 990, T H Metcalfe and P H McElroy, were invited. It is not known who drove 990 back to Walton almost certainly via former route 36.

[Above] C Hylemen collection

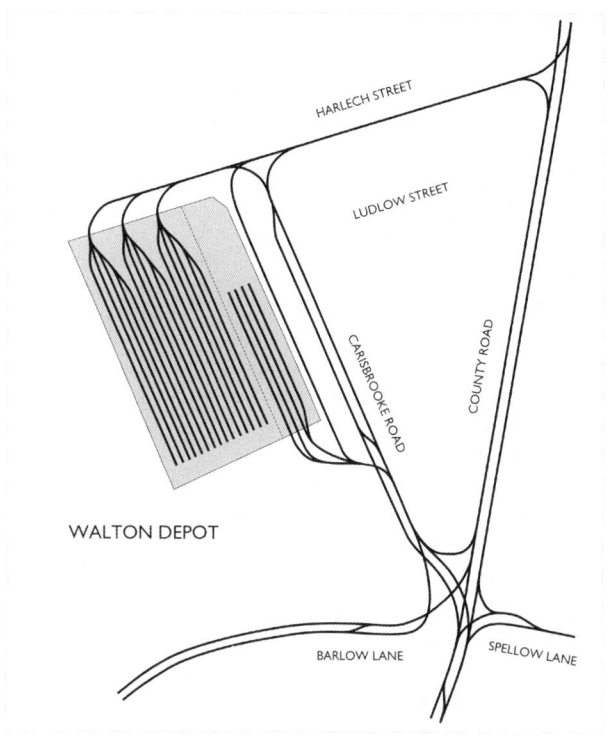

No one really mourned the passing of Bootle's trams. However, this evocative poem *The Seaforth Tram*, published by Middleton Press in *Liverpool Tramways 3: The Northern Routes* by Brian P Martin and attributed to F.J.C. provides a lasting memory.

It was midsummer's eve in Bootle,
The rain was pouring down,
The soot lay dark in Derby Park,
The fog hung over the town.

Down at Bankhall tram stop,
Eeh! There was a jam.
There were hundreds of folk
All of them soaked,
Awaiting the Seaforth tram.

'Eeh! Mother. Hark at thunder.'
Said little one to his Ma.
'That ain't no thunder,
You soft-headed Dunder.
That's only the Seaforth car'.

They talk about buses in London.
And cycles in old Amsterdam.
But never, I feels there is 'owt on wheels
Compares with the Seaforth tram.

And though I live to be ninety,
The memory's burned in my brain.
Of the earth-shaking sound
Of a tram Seaforth-bound,
And the drip of the Bootle rain.

Following these conversions, the former horse car depot at Walton was probably closed although it may have stored cars awaiting scrap for several more months. Latterly, its four tracks had housed a mix of Standards together with snowploughs and other work equipment. Access involved much manoeuvring especially when a car was moved from the main shed. The shunter left the new depot via Harlech Street, reversed and then drove the car onto the paved siding adjacent to the old building and parallel to the tracks on Carisbrooke Road. "The practice was for trams to run down this siding and reverse into the old shed until such times as it was full. After that, the cars would be stabled, bumper to bumper on the siding itself." (J A Bryant) As such activities often took place at night, nearby residents endured many a disturbed night. [Above] Martin Jenkins collection/Online Transport Archive

1951 | 117

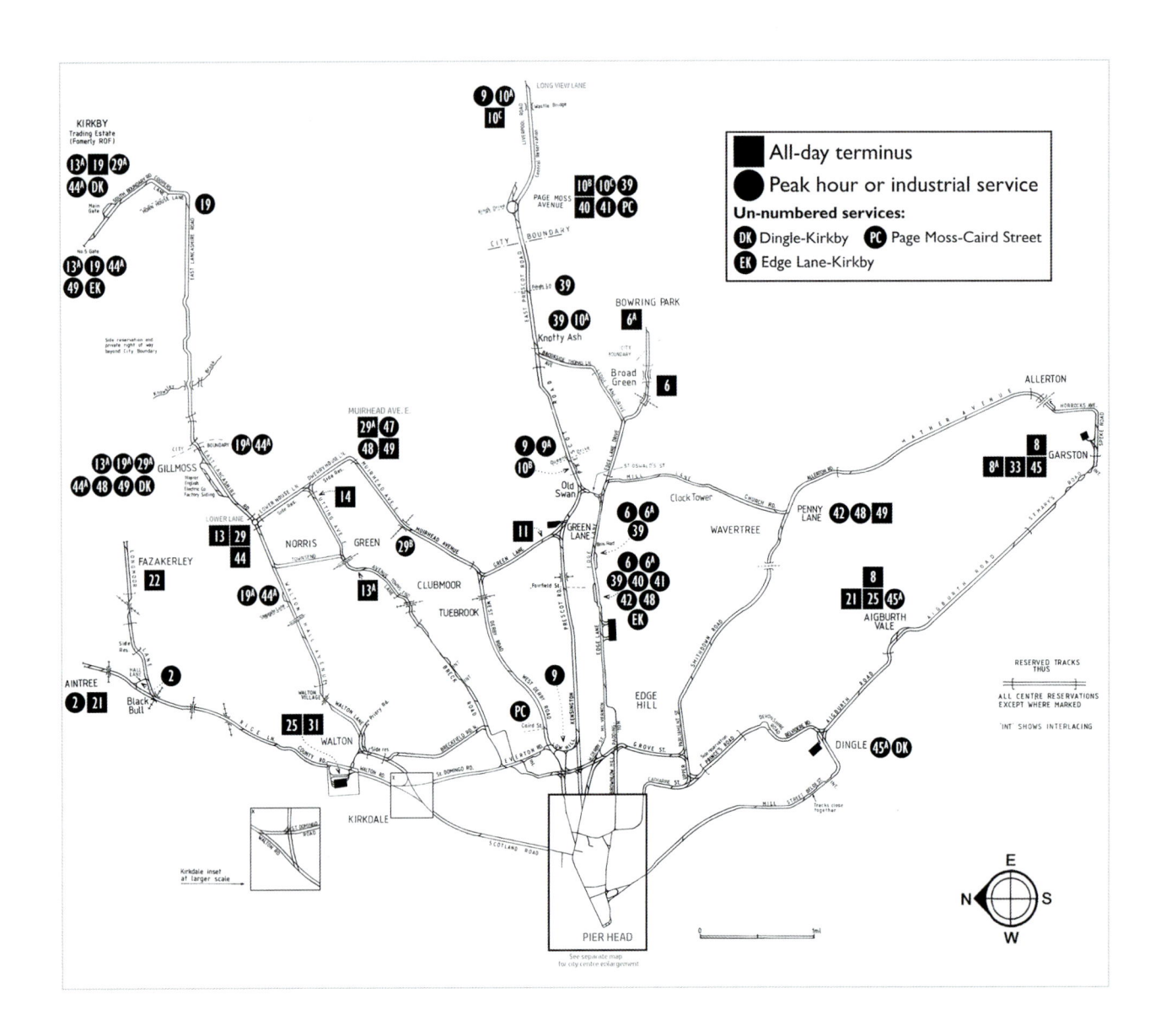

February 1951: Reconditioned standard 747 given a thorough overhaul.

March 1951: Despite recent improvements in the overall condition of the fleet, the LTPA suspect there was still a policy of minimum maintenance. They logged their concerns under a series of headings.

Maintenance:
Loose and creaking bodies. Drooping platforms. Missing ventilators. Rattling windows – won't close or are permanently closed. Faulty doors on older cars. General cleanliness.

Trucks:
Coil springs missing from Radial Arm type. Loose springs. U-bolts dropping off. Bent main frames. Track brake shoes loose. Lifeguards tied up with rope.

Overhaul and partial overhauls:
Overhauled trams emerge from works with loose bodies, faulty doors, vents bent and panels in wrong order. Trucks not touched so cars often back into works. No system of overhaul – during 1950, solid cars scrapped and loose 'wrecks' soldier on, eg 137, 577 and 694.

Cars painted twice! – 168, 169, 769, 775, 792, 866, 929.

Cars out of service for over a year:
161 (since 1948), 180 (1949), 175 (1949), 981 (1948), 211 (1949), 244 (1947).

Track:
Appalling Corrugation. Broken rail. Loose joints. Soil on reserved tracks.

Depots:
"Maintenance at Dingle (excellent), Garston (good), Green Lane and Walton (bad). Too many defects on Green Lane cars – some running repairs actually carried out in the street holding up traffic" (JWG)

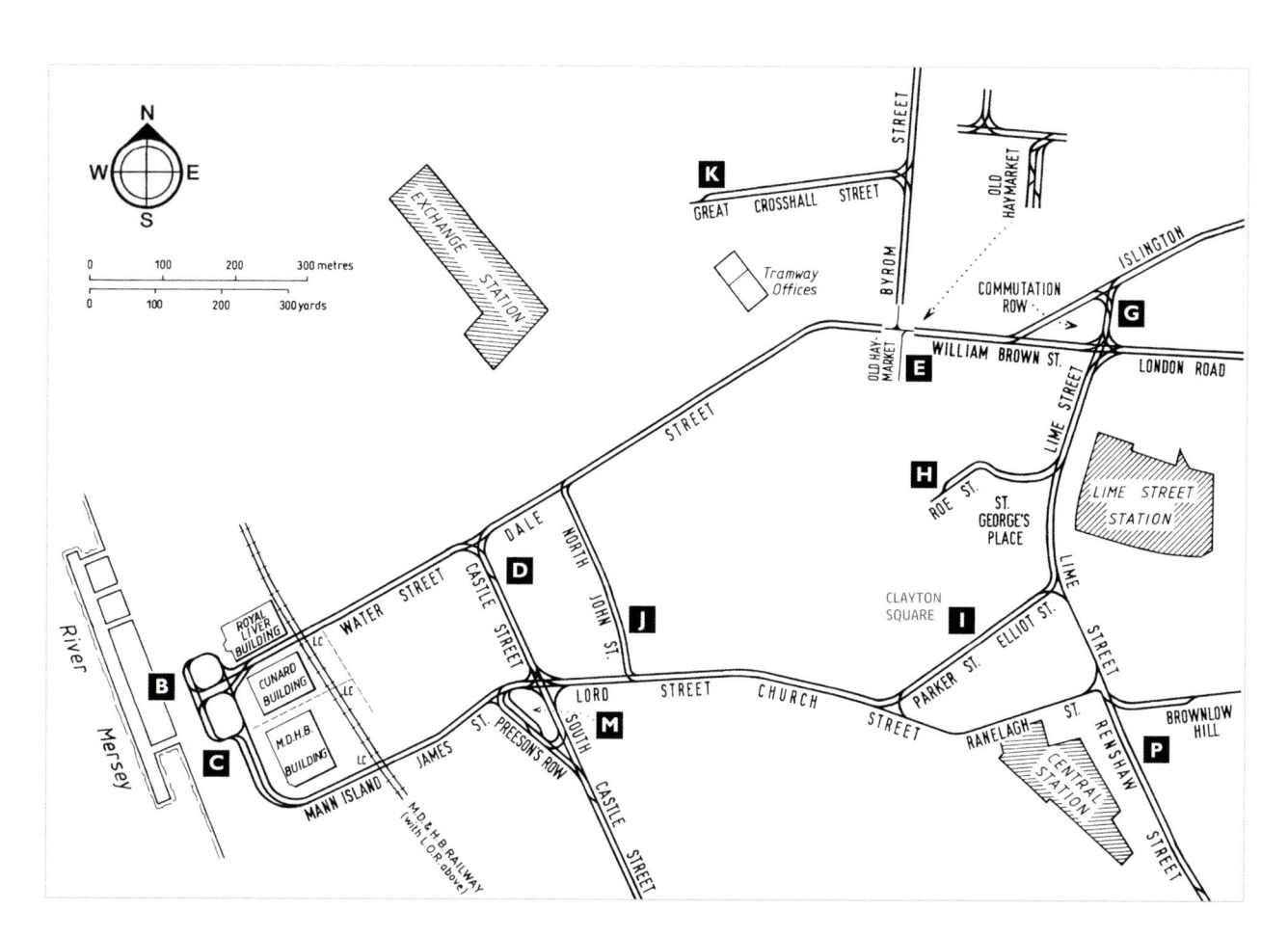

Key	Location	All day terminus	Terminal or start point for part-time routes
B	Pier Head (Centre Loop)	6 6A 10B 10C 22 29 29A 31 40	2 9 10A 29B 39
C	Pier Head (South Loop)	8 8A 13 14 19 19A 33 44 44A 45	13A 14A 45A
D	Castle Street		13A 14A
E	Old Haymarket		13 13A
G	Commutation Row *		9 10A 10B 10C
H	Roe Street *		14 14A
I	Clayton Square		10A 19A 39
J	North John Street	11	13A 14A 29A
K	Great Crosshall Street		2 22 44 44A
M	South Castle Street		21 45A
P	Renshaw Street*		8A

* Cars for these points usually showed 'Lime Street' as the destination

18 March 1951: Use of bell punch type tickets comes to an end although some were retained for emergency and pre-paid purposes. Depending upon their depot, tram conductors were issued with either TIM or Ultimate style ticket machines both of which were more efficient and led to a reduction in clerical staff. Furthermore, steps were put in place to reduce the amount of fare evasion. Tram and bus fares were also aligned.

March 1951: At some unknown date, the Dingle to Stopgate Lane industrial journeys via the 25 and 19 were discontinued.

7 April 1951: This was the last time trams were involved in the Grand National transport arrangements. Routes 21, 22 and 25 were augmented in the morning and evening and, from 10am, 36 cars maintained two shuttles, one from the Rotunda and one from Carisbrooke Road. In the evening, some 60 cars loaded at Warbreck Moor for the city, whilst the 21 was curtailed at Hall Lane. 250,000 people had watched the steeplechase.

Jack Gahan noted the following cars in use: 6, 12, 22, 30, 67, 81, 84, 95, 104, 128, 131, 152, 239, 268, 310, 313, 324, 326, 359, 380, 410, 645, 650, 677, 684, 693, 710, 719, 729, 734, 748, 751, 754, 767, 774, 777, 779, 782, 791, 796, 821, 841, 843, 852, 856, 857, 858, 862, 874, 878, 884, 901 and 979. Some, such as 6, 22, 84 and 645, were scrapped shortly after.

20 April 1951: English Electric bogie car 762 and Baby Grand 272 collided at the junction of Byrom Street and Great Crosshall Street. Both were repaired.

30 April 1951: Withdrawal of the last ex-Double Staircase 'Big Emma', although another source records 19 April. Dating from 1914, 592 had two 40hp motors, a Brill 21E truck and a post-war vestibule fitted on 22 May 1947. The lower deck sported a decorated ceiling and elaborate mouldings. It was scrapped on 23 July 1951. Here, the 66-seater is at Walton preparing to turn from Harlech Street into Carisbrooke Road where, after a short layover, it will return to Penny Lane on cross-city route 46.

[Above] N N Forbes/National Tramway Museum

May 1951: Whilst the riverside track on the Centre Loop at Pier Head was relaid and the North Loop reorganised for buses, some routes such as the 2 and 22 may have transferred temporarily to the South Loop.

Mid-May 1951: Les Folkard spent several days in Liverpool when he noted cars he rode as well as those seen in depots. Some such as 85, 89 and 826 are recorded at two different depots. He also lists The Pool at Edge Lane. This is very much a detailed personal record as he was not able to see every car in stock at the time (see table, opposite).

Since being moved from Garston, German trailer 429 and Bellamy 558, which had been acquired for preservation by the

120 | THE LEAVING OF LIVERPOOL

Edge Lane	6, 11, 14, 30, 62, 85, 89, 98, 147, 153, 177, 178, 181, 204-206, 208, 210, 219, 221, 240, 241, 245, 248-250, 253, 254, 260, 263, 275, 280, 284, 298, 321, 344, 353, 502, 555, 659, 719, 725, 726, 750, 759-761, 764, 767, 769, 783, 793, 812, 813, 964, 986
Dingle	69, 95, 152, 160, 203, 215, 226, 313, 389, 410, 440, 462, 647, 729, 754, 762, 766, 775, 776, 817, 821, 826, 827, 866, 874, 877-880, 884, 901, 905, 949, 952, 967, 969, 979, 992
Green Lane	5, 12, 29, 35, 67, 85, 89, 97, 98, 104, 168, 169, 232, 240, 289, 293, 299, 328, 338, 343, 450, 459, 673, 677, 684, 687, 688, 701, 725, 727, 732, 747, 755, 764, 772, 792, 798, 800, 803, 806, 807, 875, 886, 887, 893, 897, 904, 906, 907, 946, 950, 985, 988, 990
Walton	15, 24, 76, 81, 86, 91, 92, 95, 122, 128, 133, 140, 151, 154, 156, 158, 182, 185-188, 215, 216, 218, 222, 224, 226, 227, 238, 239, 257, 266, 268, 270, 272, 277, 278, 285-288, 297, 327, 342, 553, 566, 648, 652, 663, 676, 678, 679, 703, 722, 729, 748, 749, 751, 754, 765, 770, 771, 774, 777-779, 782, 784, 785, 788, 791, 793-796, 799, 811, 815, 817, 821, 823, 824, 826, 829, 833, 835, 836, 839, 840, 843, 845, 846, 852, 853, 855-860, 862, 864, 867, 900, 954, 963, 970
Garston	183, 184, 279, 716, 873, 881, 903, 906, 918-942, 947, 962, 968
Edge Lane Works	Paint shop: 165, 252, 790, 841, 869, 885, 914 Under repair: 160, 243, 244, 264, 380, 645, 747, 816, 828, 837, 868, 881, 889 (rebuilt on lightweight trucks), 890, 930, 951, 957. Awaiting scrap: 22, 26, 36, 38, 49, 50, 84, 90, 101, 114, 118, 144, 303, 309, 325, 359, 400, 592, 599, 637, 667, 697, 712, 724, 745, 746, 781, 787, 801, 808, 810, 814, 819, 822, 825, 831, 834, 838, 842, 861 For preservation: 429, 558

LRTL Museum Committee, had been stored on the scrap lines. 558 had survived as a war-time cloakroom for conductresses and still had a 'Ladies Only' sign in the saloon window. 429 had become a tool store known as 'Arthur's old van'. The second view shows it sandwiched between Standard 22 and former snowplough 122. Note the pre-war tower wagon on the right.

[Right, top, centre] Phil Tatt/Online Transport Archive (both)

Also on the dump was Marks Bogie 819 which had just been withdrawn, Jack Gahan describing it as 'in good condition'. Being non-standard it was probably an early casualty having been fitted in 1937 with K3 controllers and Blackpool-style English Electric trucks, which can clearly be seen in this view at Dingle.

[Right, bottom] Martin Jenkins collection/Online Transport Archive

May-June 1951: The life-expired junction at Edge Lane/St Oswald's Street was replaced by a roundabout circled by tracks as well as a new segregated approach on the west side. On 30 May 1951, 260 negotiates the temporary track laid over the new rail circumnavigating the roundabout. In the background is one of the tower wagons. Once the revised layout was complete, short-workings formerly turning on Springfield Road crossover now circled the roundabout. *[Overleaf, top] Liverpool City Engineers/ National Tramway Museum collection*

24 June 1951: Annual LRTL tour held on recently reconditioned 914. Each year, a carefully planned itinerary meant participants visited sections earmarked for closure. Departing from Roe Street in early afternoon, there was a high speed run to Garston via the 6, 49 and 8 from where the car proceeded to Kirkby via the 33, 25, 13 and 19. Later, visits were made to Pagemoss and Fazakerley where 914 waits for three service cars to reverse. *[Below] F N T Lloyd-Jones/Online Transport Archive*

26 June 1951: Dingle cars 871, 878, 884, 979 and 992 sent to Garston and 976 to Walton.

13 July 1951: Night-time fire caused by a smouldering cigarette on board 140 inside Walton depot. The badly scorched car, with most of its windows broken, quickly driven out onto the street.

4 August 1951: Route 21 discontinued. This closure, and that of the 45, were opposed by the Liverpool Overhead Railway who feared the replacing buses would intensify competition. Operated mainly from Dingle, this was the longest of the regular cross-city routes and was almost entirely on street track. It linked Aintree in the north with Aigburth Vale in the south via Warbreck Moor, Walton Vale, Rice Lane, County Road, Walton Road, Kirkdale Road, Scotland Road, Byrom Street, Dale Street, Castle Street, South Castle Street, Canning Place, Park Lane, Mill Street, Beloe Street, Dingle Mount and Aigburth Road. The Monday to Saturday headway was every eight minutes dropping to every 12 on Sundays. At peak times, extras ran in both directions from South Castle Street. Other known short workings were to Rotunda and Dingle. The queuing point on Nimrod Street, where Everton supporters waited for post-match extras heading south, was discontinued at the end of the 1950/51 season.

"I rode this route just before it closed by which time I had graduated to recording fleet as well as route numbers. I took a 45 to Aigburth and was fascinated when 958 went down steeply-graded Beloe Street with its interlaced and closely spaced double track on which cars were unable to pass. I remember we had to wait for the lights to clear until another car passed in the opposite direction. I heard later that the City Engineers adopted this layout to save on costly point work. At Aigburth, I watched a few 21s and 25s using the trolley reverser before boarding 313 for the 8-mile run to Aintree. People hopping on and off all the time with especially heavy loads after leaving Dale Street. 313 was in good nick and bustled along at a fair speed. As I wanted to ride a Standard with a 'pre-fab' windscreen, I waited for 734 which I knew was somewhere behind as it we had passed it on the way. This definitely offered a rougher ride than 313. Off at Walton for a quick look at the depot. Saw a handful of cars occupying the siding on Carisbrooke Road. Boarded 664 for the run back to Aigburth on the 21. Another chance to sample the Beloe Street incline. Back on a 45 so I could meet my mother beside the Victoria Monument." **(MJ)**

Route 22 from Fazakerley to the Pier Head was replaced by buses. Another heavy loader, it was operated from Walton depot. After using the trolley reverser at Fazakerley, cars ran along Longmoor Lane (12mph speed restriction) to Black Bull from where they paralleled the 21 as far as the Town Hall after which they proceeded to the Pier Head via Water Street. Short workings included peak hour departures from Great Crosshall Street as well as a few journeys to and from Rotunda, Black Bull (Hall Lane) and Walton depot. Owing to the high volume of traffic, especially along Scotland Road and Walton Road, the 22 ran every seven minutes except on Sunday mornings when it dropped to every 10 minutes.

5 August 1951: Although the 2 was now an all-day service from Pier Head to Aintree, in peak hours additional cars still ran to and from Great Crosshall Street and Black Bull (Hall Lane) and possibly Walton Church and Warbreck Moor. As there were insufficient buses to handle the volume of traffic, the all-day 2 compensated for the loss of the 21/22 on the Scotland Road/Walton Road corridor. Also, by retaining a presence deep in Ribble territory, the Corporation was able to replace the route with its own buses.

6 August 1951: To help offset the loss of the southern portion of the 21, a new, short-lived service 45A was introduced. Operated from Dingle, it lasted just a month and ran between Dingle and South Castle Street. However, this number had been used for many years to cover a variety of short-workings on the 45.

July/August 1951: On the bank: 88, 310, 359, 450, 645, 650, 716, 718, 746, 749, 780, 822, 832, 838 and 981. Most sold for scrap in September.

122 | THE LEAVING OF LIVERPOOL

6 September 1951: City still plagued by occasional power-cuts some of which lasted for several hours. For example, on this damp Monday evening no trams could move between 5.20pm and 8pm. *[Above, both] Leo Quinn collection/Online Transport Archive*

8 September 1951: All day service 45 (Pier Head-Garston) and part-day service 45A (South Castle Street-Dingle) operated for the last time. Serving the southern suburbs, inbound 45s arrived at the Pier Head, South Loop via James Street and Mann Island but departed via Water Street and Castle Street. Paralleling the 45A from South Castle Street, both covered older, densely-populated inner suburbs via Canning Place, Park Lane, Mill Street, Beloe Street and Dingle Mount as far as Dingle where the 45A terminated whilst 45s continued to Garston via Aigburth Road and St Mary's Road. In the early mornings, some 45s departed from Hill Street (near Mill Street) and in the evening peak a few 45s short worked to either Dingle or Aigburth. The headway on the 45 was every 20 minutes Monday to Friday but only every 30 minutes on Saturdays and Sundays when operation did not begin until mid-day. Time-table information relating to the supplementary 45A has not come to light but it is known to have been reasonably frequent on Saturdays.

ROUTE PROFILE
21, 22

Aintree. *G W Price Collection*

Longmoor Lane, central reservation. *F NT Lloyd-Jones/Online Transport Archive*

Single track over Barlow's Lane Bridge, Fazakerley station. *F NT Lloyd-Jones/Online Transport Archive*

Longmoor Lane, roadside reservation. *J H Roberts/Online Transport Archive*

Rice Lane. *A H Jacob*

LIVERPOOL CORPORATION PASSENGER TRANSPORT

TRAM ROUTES

AIGBURTH **21** AINTREE

FAZAKERLEY **22** PIER HEAD

ALTERATIONS AND
CONVERSION TO BUSES

Commencing
SUNDAY, 5th AUGUST, 1951

124 | THE LEAVING OF LIVERPOOL

Byrom Street. *Martin Jenkins collection/Online Transport Archive*

Outside Town Hall. By July 1951, 664 of 1921 was the oldest car in service. *Martin Jenkins collection/Online Transport Archive*

Dingle Mount. *N N Forbes/National Tramway Museum*

Beloe Street
[Above] F N T Lloyd-Jones/Online Transport Archive

Aigburth Vale. *J W Gahan*

Aigburth Vale, last day. *[Left] Leo Quinn Collection/Online Transport Archive*

Journey time	55 minutes (21), 44 minutes (22)
Last day of operation	4 August 1951
Cars seen in service on last day	167, 310, 958
Last cars	310 (21, northbound), 958 (21, southbound); last 22 not known
Track abandoned	Most of Longmoor Lane (1¼ miles of mainly centre and roadside reserved track)

1951 | 125

ROUTE PROFILE
45, 45A

South Castle Street. *A S Clayton/Online Transport Archive*

Passing Sailors' Home, Canning Place. *N N Forbes/National Tramway Museum*

Mill Street, junction with Park Road where the tracks last used in August 1949 were still in situ. *A S Clayton/Online Transport Archive*

Mill Street. *A S Clayton/Online Transport Archive*

Beloe Street. Movements on and off this intriguing section of track were controlled by light signals (inset). *A S Clayton/Online Transport Archive*

126 | THE LEAVING OF LIVERPOOL

Dingle Mount. *A S Clayton/Online Transport Archive*

Dingle. *Alan B Cross*

Garston. *F N T Lloyd-Jones/Online Transport Archive*

Allerton bridge. *Roy Brook/National Tramway Museum*

LIVERPOOL CORPORATION PASSENGER TRANSPORT

TRAM ROUTE

GARSTON 45 CITY

ALTERATIONS AND CONVERSION TO BUSES

Commencing
SUNDAY, 9th SEPTEMBER, 1951

Journey time	33 minutes (45), 22 minutes (45A)
Last day of operation	8 September 1951
Cars seen in service on last day	251, 901, 919, 926, 935, 951, 958.
Last cars	Not known
Track abandoned	Castle Street, Canning Place, Park Lane, Mill Street, Beloe Street, Dingle Mount.

1951 | 127

Castle Street: This elegant central area thoroughfare was used for the last time by routes 6, 6A, 14, 14A, 19, 19A, 19B and outbound 45. This view, taken earlier in the year, shows the whole length of the street as far as the Town Hall. The wiring was quickly removed, the street repaved and the worn junctions at either end replaced by new straight track. *Leo Quinn collection/Online Transport Archive*

South Castle Street: Known to generations of staff as 'Castle' or 'The Castle' this was now no longer a regular tram terminus. Some points and crossings were lifted and other sections covered over but a single clockwise circuit was retained for emergencies. Oddly, it was also used for one journey a week – a Sunday duty to Broadgreen Hospital which still turned at South Castle Street although the destination was not shown on the car's indicators.

Dingle depot: With replacement of the 45s, the depot closed to trams although emergency access was retained until at least January 1952. The 25 was transferred to Walton and the Dingle-Kirkby to either Walton or Edge Lane, or possibly both. The staff had an excellent reputation for looking after its cars and it is said successive depot foremen tried to stop their allocation going for attention at the Works. The depot also housed the training schools for tram drivers and conductors. After serving as a bus garage it closed in 1965 and was demolished in 1993.

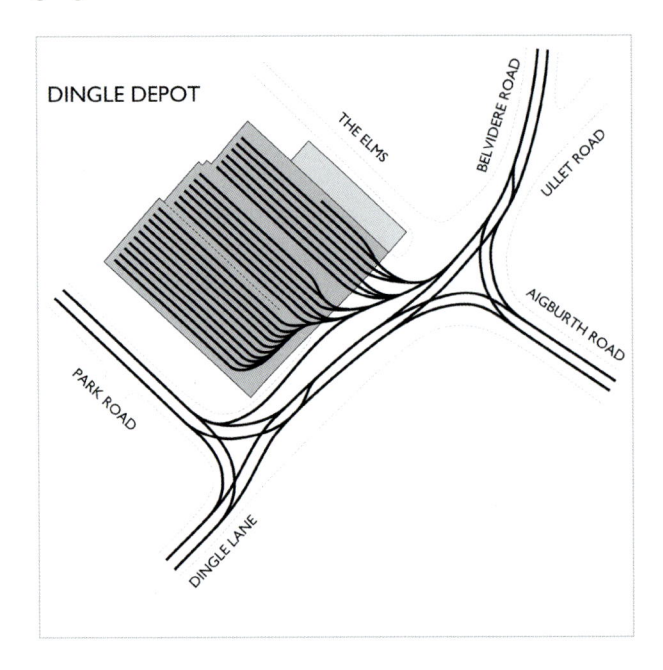

Stan Watkins logged the following car movements:

Dingle to Garston	160, 162
Dingle to Walton	152, 165, 877, 879, 880, 901, 905, 945, 949, 972
Dingle to Edge Lane	952, 958
Garston to Edge Lane	203, 207, 247, 258, 267, 269, 271, 273, 279, 283, 947, 974

9 September 1951: With the closure of Castle Street, routes 6, 6A, 39 and 40 transferred to the Pier Head, South Loop whilst the 14s and 19s were rerouted via James Street and Mann Island. From the same date, the loading point for the 8s was probably relocated to the siding on the south side. The Centre Loop was used by the 2, 10B, 10C, 29 and 29A plus occasional 9, 10A and 29B.

At first redundant track was lifted including some 1400 tons during 1950/51. However, the work was disruptive and labour-intensive and the cost of resurfacing approximately £30,000 per mile. So when the value of scrap metal dropped, rails were simply left in situ or covered over. However, where the entire road needed renewing, as on Tithebarn Street (seen here), track continued to be removed. *[Opposite, top] F N T Lloyd-Jones/Ovnline Transport Archive*

September 1951: Transport Committee approved the scrapping of modern cars held pending possible repair: 805 (withdrawn since 1949), 808 (collision 1950), 814 (collision December 1949), 822 (withdrawn 1950), 831 (collision 1948), 861 (withdrawn June 1950) and 832 (broken frame 1949). On the credit side, 957, 965, 973, 974, 975, 978 and 983 were rehabilitated after laying up since 1949/50.

15 September 1951: Since conversion of the 21/22, the surviving handbrake Standards rarely appeared outside peak hours. However, Jack Gahan noted Green Lane's 344 on all-day duty on the 29A until 6.30pm. Two months later it was withdrawn. His brother Ted recalled "When I worked split turns at Green Lane in the early 1950s, I would often be assigned an older Edge Lane car to work a couple of trips to and from Pagemoss or possibly on the 'bottom road' as the 29s were known"

A few Standards were even repainted during 1950/51 with a cream band below the upper deck windows. Some, such as 459, 726 and 727, had the city coat of arms on the waist panel but this is known to have been omitted on 701, 732 and 747. 726 is at Lower Lane in the summer of 1951. *[Opposite, bottom left] Phil Tatt/Online Transport Archive*

The following airbrake Standards cars are known to have been repainted during their final years but omitting the cream band below the upper deck windows: 5, 12, 35, 81, 87, 89, 95, 97, 98, 147, 328, 338, 751 and 753. The side indicators on most had fallen into disuse and many were removed. 97 is seen on Muirhead Avenue in mid-June 1952 working a peak hour 48 from Gillmoss to Penny Lane. *[Opposite, bottom right] Peter Mitchell*

1951

ROUTE PROFILE
2

Peak hour turn-back at Hall Lane, Black Bull in 1942. *N N Forbes/National Tramway Museum*

Walton Vale/Orrell Lane junction. *N N Forbes/National Tramway Museum*

Black Bull, Rice Lane/Longmoor Lane junction. *N N Forbes/National Tramway Museum*

Rice Lane. *N N Forbes/National Tramway Museum*

Walton Church. *N N Forbes/National Tramway Museum*

Pier Head Centre Loop. *Julian Thompson/Online Transport Archive*

Journey time	40 minutes
Last day of operation	10 November 1951
Cars seen in service on last day	214, 772, 783, 816, 866
Last cars	Not known
Track abandoned	Walton Road (beyond Harlech Street), County Road, Rice Lane, Walton Vale, Hall Lane, a short stretch of Longmoor Lane, Warbreck Moor.

1951 | 131

October 1951: Norman Forbes paid for LRTL Museum Committee-owned 429 and 558 to be taken by low loader into open storage at Kirkby. No photographs of this move have come to light.

10 November 1951: The short-lived all-day route 2 replaced by buses. From Pier Head, cars reached Aintree via Water Street, Dale Street, Byrom Street, Scotland Road, Kirkdale Road, Walton Road, County Road, Rice Lane, Walton Vale and Warbreck Moor. At peak hours, there were additional departures from Great Crosshall Street. Other known short-workings included Walton Church, Black Bull (Hall Lane) and Warbreck Moor as well as depot only journeys. A 10 minute service was provided Monday to Saturday except in the early morning and late evening when it dropped to 20 minutes. On Sundays, frequencies gradually increased from every 45 to every 13 minutes. This all-street track route was the last to thread its way through many densely-populated, close-knit north end communities.

> "Quite a slow ride with increasing congestion from cars and lorries especially on some of the narrower sections in the Walton area. Rows of small shops and streets of terraced houses. Passed close to Walton Gaol and Goodison Park, home of Everton FC. Hard now to remember the lines of trams packed nose to tail at Aintree on race days when the tram seemed king. Terminus almost deserted as I watched 791 use the trolley reverser. My last ride beyond Walton." **(MJ)**

With replacement of the 2, another of the rapidly dwindling cross-city transfer points at Spellow Lane was eliminated with the remainder set to disappear during 1952.

Christmas 1951: No handbrake Standards assisted with the pre-Christmas rush. During the period covered by this book, trams ran on Christmas Day and Boxing Day to either a 'special' or Sunday schedule. Certain industrial services also ran 'as required'.

31 December 1951: 346 cars officially in stock for the 15 remaining basic services.

Cars believed to have been reconditioned during 1951 some with internal body bracing: 165, 166, 178, 186, 204, 211, 219, 222, 229, 240, 244, 249, 257, 271, 289, 871, 914, 945, 957, 965, 973-975, 978 and 983 with others undergoing major bodywork.

Cars believed withdrawn during 1951: As in previous years, the precise withdrawal and breaking up dates are not always known.

Month	Cars
February	65 (laid up since April 1949), 68, 368, 701
March	49, 90, 114, 353 (broken up 19 March), 598, 696
April	26, 38, 61, 84, 118, 592, 667, 712, 819
May	6, 109, 110, 645, 697, 702, 732
June	679
July	11, 22, 36, 87, 101, 144, 301, 359, 639, 680, 722
August	321, 637, 663
September	69, 77, 309, 326, 373, 674, 676, 678, 683, 695, 710, 781, 787, 805, 808, 810, 814, 822, 825, 831, 832, 838, 842, 861
October	14, 18, 29, 67, 131, 140, 389, 410, 688
November	15, 76, 85, 86, 92, 104, 132, 327, 650, 675, 677, 687, 734, 739 (only Standard with large indicator boxes)
December	450

Other cars withdrawn during the year but month unknown: 24, 40, 50, 88, 99, 130, 133, 135, 324, 325, 342, 376, 380, 441, 647, 659, 685, 716, 718, 724, 729, 748, 749, 750.

In addition 429 and 558 were sold to the LRTL Museum Committee, as described earlier.

CHAPTER 9 | 1952

For the first time since the war the tramway was in reasonably good overall condition with miles of recently relaid track and scores of repaired and repainted cars, the majority of which were under 20 years old and in reasonable structural, mechanical and electrical order. As a result, reliability had also improved.

> "All cars in service are in a clean presentable condition. A great deal of track has been relaid and the amount of single track (apart from one-way streets) has been reduced to a negligible quantity. Unfortunately, there is a great deal of corrugation which does harm to the reputation of the tramway." **(Norman Forbes)**

Stage Two of the conversion programme led to closure of miles of recently relaid track, much on reservation.

Tram Route Stage No.	KIRKBY—DINGLE	Stage No.
57	Kirkby Estate	—
59	1½ Ormskirk Road	76
61	2½ 1½ Ainsworth Lane	74
63	4 2½ 1½ Radshaw Nook	72
65	4 4 2½ 1½ Gillmoss (Stonebridge L./Back Gillmoss L.)	70
67	5 4 4 2½ 1½ Lower Lane	68
69	5 5 4 4 2½ 1½ Oak Lane	66
71	6 5 5 4 4 2½ 1½ Lewisham Road	64
73	6 6 5 5 4 4 2½ 1½ Queens Drive	60
75	6 6 6 5 5 4 4 2½ 1½ Tuebrook Station	58
77	6 6 6 6 5 5 4 4 2½ 1½ Sheil Road	56
79	7 6 6 6 6 5 5 4 4 2½ 1½ Everton Road	54
81	7 7 6 6 6 6 5 5 4 4 2½ 1½ Pembroke Place	52
83	7 7 7 6 6 6 6 5 5 4 4 2½ 1½ Myrtle Street	50
85	7 7 7 7 6 6 6 6 5 5 4 4 2½ 1½ Nth. Hill St.	48
—	7 7 7 7 7 6 6 6 6 5 5 4 4 2½ 1½ Dingle	46

4 January 1952: Last day of the unnumbered cross-city industrial service from Kirkby or Gillmoss to Dingle. Operating via routes 19/29/25, southbound cars sometimes showed 25 and northbound 29 although latterly they often ran with blank number screens.

Walton-based 770 has a reasonable load as it waits to leave the Admin loop at Kirkby for Dingle in September 1951. 790 follows with a short-working 44A. *N N Forbes/National Tramway Museum*

This view of 762 heading for Dingle was taken at 6pm at Muirhead Avenue East and was probably the 5.40pm departure from Kirkby – the other southbound journey leaving Gillmoss at 5.40pm. 762 was then based at Edge Lane which seems to suggest the 11-mile route was probably operated latterly from Edge Lane and Walton depots. However, this conjecture raises several unanswered questions: for example, did the car working the 6.55am from Dingle come from Edge Lane? What happened after it arrived at Kirkby? Did it return to base or cover other duties until one of the scheduled evening returns? Did the car covering the additional evening departure from Gillmoss come from Walton depot and then return from Dingle to Walton or could it have been housed overnight at Dingle? *D Roberts*

ROUTE PROFILE
25

Carisbrooke Road. Note the wiring into the former horse car shed has been removed. *A S Clayton/Online Transport Archive*

Walton Road. *A S Clayton/Online Transport Archive*

Everton Road. *A S Clayton/Online Transport Archive*

Moss Street/London Road. *A S Clayton/Online Transport Archive*

Grove Street/Upper Parliament Street. *A S Clayton/Online Transport Archive*

Princes Road. *A S Clayton/Online Transport Archive*

134 | THE LEAVING OF LIVERPOOL

Aigburth Vale. Note the trolley reverser. *A S Clayton/Online Transport Archive*

Tram Route 25	WALTON—AIGBURTH VALE
Stage No.	Stage No.

65					Spellow Lane			—
67	1½				Royal Street			68
71	2½	1½			Breck Road			62
73	4	2½	1½		Pembroke Place			60
75	4	4	2½	1½	Myrtle Street			58
77	5	4	4	2½	1½	North Hill Street		56
79	6	5	4	4	2½	1½	Dingle	54
81	6	6	5	4	4	2½	1½ Lark Lane or Elsmere Avenue	52
—	6	6	5	5	4	4	2½ 1½ Aigburth Vale	50

Routes 25, 31, DINGLE-KIRKBY
—CONVERSION TO BUSES

Commencing SUNDAY, 6th JANUARY, 1952

ROUTE
AIGBURTH VALE **25** WALTON

ROUTE
PIER HEAD **31** WALTON

ROUTE
DINGLE—KIRKBY
NOW ROUTE **94**

Journey time	36 minutes
Last day of operation	5 January 1952
Cars seen in service on last day	152, 155, 158, 782, 791, 806, 820, 836, 873, 969, 975
Last car	873 (Walton-Aigburth), 975 (Aigburth-Walton)
Track abandoned	Parts of Grove Street although the cross-over at the south end retained for emergencies for a short time, possibly until June 1953.

ROUTE PROFILE
31

In its final manifestation, the all-street-track 31 duplicated the 25 from Walton to Moss Street from where it reached the Pier Head via London Road and Dale Street. Operated from Walton depot, only two cars were needed for the 30 minute Monday-Sunday off-peak headway. The 31 provided a useful link between parts of Everton and the central business district. Photographs indicate some unadvertised peak hour cars departed from Old Haymarket.

Harlech Street/Carisbrooke Road. *A D Packer*

Grant Gardens. *A S Clayton/Online Transport Archive*

Moss Street, turning into Brunswick Road. *A S Clayton/Online Transport Archive*

Pier Head, South Loop. *G W Price collection*

Journey time	30 minutes
Last day of operation	5 January 1952
Cars seen in service on last day	779, 823
Last car	779
Track abandoned	None

136 | THE LEAVING OF LIVERPOOL

5 January 1952: Final day of operation of Routes 25 and 31. The 25 linked Walton (Carisbrooke Road) with Aigburth Vale by way of Walton Road, St Domingo Road, Heyworth Street, Everton Road, Brunswick Road/Erskine Street, Moss Street, Crown Street, Grove Street, Princes Road, Belvidere Road and Aigburth Road. Except for a 20 minute headway on Sunday mornings, it operated every 10 or 12 minutes but with extras in peak hours. Worked from Walton, it provided an important cross-city connection. Few people made the full journey as many transferred to other routes. In the late evenings, 25s were frequently packed with people patronising cinemas in the London Road area. No known short workings although inspectors may have turned cars at Dingle. This ended regular use of Aigburth Vale's trolley reverser, the last on the system, which was removed at some unknown date.

6 January 1952: Last handbrake Standards officially withdrawn including 30, 35 (collision), 62, 302, 462, 646, 648, 652, 678, 682, 684, 700, 703 and 719, of which 30, 646, 684 and 703 were converted into snowploughs ready for the winter months. Also withdrawn was air-brake Standard 313. Cars numbered between 770 and 867 all at Walton depot.

6 January 1952: Allocation lists for two depots have survived

Edge Lane	5, 12, 89, 97, 98, 147, 153, 170, 172, 174, 177-179, 181, 201, 202, 204-206, 208, 210-216, 218-224, 226, 230, 235-237, 241, 243-250, 252-254, 260, 261, 263, 264, 274-276, 280, 284, 296, 440, 723, 725, 726, 732, 759-762, 764-767, 769, 917, 945, 948, 957, 958, 964, 978. (Total 82)
Garston	160, 162, 183, 184, 869, 871-874, 878, 881, 883, 884, 903, 918-942, 949, 955, 962, 966-968, 972, 976, 979, 984, 992. (Total 50)

28 January 1952: During the winter of 1951/2, some Bellamy snowploughs were in action. 502 is hard at work near Broadgreen Station. Most drivers donned heavy coats, hats, mufflers, boots, thick gloves and, sometimes, goggles although visibility must have been virtually nil in driving sleet and snow. *[Right, top] Martin Jenkins collection/Online Transport Archive*

1 April 1952: There were now just over 60 route miles with an operational fleet of 331 cars including 24 Reconditioned air-brake Standards.

February-May 1952: It is believed the following cars were withdrawn: 89, 673, 726, 729, 745, 746, 753, 755, 776 and 821.

30 May to 2 June 1952: LRTL hold their annual convention in Liverpool. On Saturday 31 May, members were shown round the Works by R J Heathman, the Rolling Stock and Works Engineer. Formerly employed by English Electric, he had designed new cars for places like Belfast, Rotherham, Leeds and Sunderland. At Liverpool, he was heavily involved in the design of the Liners. He told the party there were 320 modern trams and, as it was planned to keep some for another six years or so, the space in the Works was to be divided into tram and bus sections. Everyone was impressed with the smart appearance of most of the fleet including freshly-painted 243 and 242. *[Right, centre] J H Price/National Tramway Museum*

On the Sunday, a 47-mile tour took place on board newly-painted Maley 942. Amongst those on board were J W Fowler, Chairman of the LRTL and C Busch, a native of Melbourne. The driver was Inspector Penman and the conductor LRTL member, Ted Gahan.

The 56 participants enjoyed a fast run from Roe Street to Kirkby. A short distance from 5 Gate terminus, they inspected a newly installed crossing with a branch of the internal rail system built to serve the Yorkshire Imperial Copper tube factory. At the time of the visit, wrong line working was in force on this section. *[Above, left]*
W J Wyse/LRTA (London Area)/Online Transport Archive.

After parking 942 on Admin loop, members walked to the premises of John McGeoch & Sons where they found trailer 429 and Bellamy 558 clearly suffering from exposure to the elements.
[Bottom] F N T Lloyd-Jones/Online Transport Archive

From Kirkby, the tour headed to Walton and then Longview Lane by way of Heyworth Street and Low Hill. From there, 942 went via the 40 and 6A to Bowring Park and finally to Garston via the 6A, 49 and 8. In the evening, Norman Forbes gave a presentation on the history of the tram fleet and the Overhead Railway. The next routes to close should have been the 42, 48 and 49 but these were reprieved when the rails on the reservation between Pagemoss and Longview Lane were condemned as unsafe and the conversion of routes 9, 10A and 10C was brought forward. A proposal to keep the section open during peak hours was rejected by the City Engineer. *[Above, right]* E A Gahan/Online Transport Archive

20 June 1952: This Friday marked the last day for the Caird Street to Longview Lane industrial workings, which were now cut back to Pagemoss. Also discontinued were the few evening departures from Southbank Road to Longview Lane via route 40. 343 is seen on one of these duties heading east along Edge Lane. This car entered service as No 233 in 1932. After being reconditioned in 1936, it was renumbered in 1937 and fitted for a short time with an experimental bow collector. Note the unusual ventilation above the lower deck windows. *[Below]* Peter Mitchell

138 | THE LEAVING OF LIVERPOOL

21 June 1952: Routes 9, 10A and 10C operate for the last time. The 6¾ mile 10C was the principal service, the number having only been introduced in 1939. It ran from the Pier Head, Centre Loop, to Longview Lane (sometimes abbreviated to Longview on screens) via Dale Street, Kensington, Prescot Road, Old Swan, East Prescot Road and Liverpool Road. Officially, the 9 operated Monday to Friday evenings from Commutation Row to Old Swan and the 10A from Clayton Square to Knotty Ash. In practice, both numbers were used for a variety of short-workings on the Prescot Road corridor including many additional evening departures from the peak hour queue points at North John Street, Old Haymarket, Clayton Square, Commutation Row and Low Hill. Early post-war views exist of 10Cs at Castle Street. From these points, cars worked to any of the following destinations: Green Lane depot, Old Swan, Finch Lane, Pilch Lane, Pagemoss and Longview Lane. In the mornings, extras were run from all the above to Low Hill, Clayton Square and North John Street. Sometimes cars working back to Green Lane or Edge Lane depots showed 9 or 10A. There is also a photograph of a 9 leaving Longview for Commutation Row (Lime Street on the screen). Sometimes as many as eight cars could be waiting to reverse at Longview Lane.

To handle the high demand on the Prescot Road corridor, the 10C operated every 12 minutes through most of the week including Sundays. To provide a six minute headway on the busiest section, it was supplemented by the 10B which operated every 12 minutes but only as far as Pagemoss. In rush hours, extra 10B and 10C were provided by Green Lane and Edge Lane depots. This often resulted in bunching with cars running at two minute intervals. *[Right, top] Peter Mitchell*

Considerable public disquiet followed this particular conversion. The replacing buses failed to cope, time-keeping was erratic and buses had to be commandeered from other routes. Some passengers living between Pagemoss and Knotty Ash switched to the more reliable tram route 40.

To make matters worse, the 10B was now little more than a peak hour service although four cars did maintain a 20 minute off-peak headway Monday to Saturday. However, the running time was so slack, drivers were often crawling between stops on the first couple of notches.

During the evening peak, many extra 10Bs loaded in Commutation Row but increasing congestion, caused in part by the replacement buses, led to more bunching. Furthermore, in the mornings, passengers were confused because any city-bound car showed 10B regardless of its ultimate destination. 950 passes through the suburban shopping centre at Old Swan. The body on this Liner was in poor structural condition. *[Below] Peter Mitchell*

ROUTE PROFILE
9, 10A, 10C

Dale Street. *Peter Mitchell*

North John Street, turning into Lord Street. *A B Cross*

London Road. *Peter Mitchell*

London Road. *Peter Mitchell*

Low Hill. *Peter Mitchell*

Kensington. *Peter Mitchell*

Prescot Road. *Peter Mitchell*

Prescot Road/Sheil Road. *Peter Mitchell*

140 | THE LEAVING OF LIVERPOOL

Dinas Lane, north of Brookfield Avenue. *Peter Mitchell*

Liverpool Road, east of Pagemoss. *Peter Mitchell*

Liverpool Road. *Peter Mitchell*

Approaching Longview. *E A Gahan/Online Transport Archive*

LIVERPOOL CORPORATION PASSENGER TRANSPORT

TRAM ROUTE

LONGVIEW **10c** PIER HEAD

(And the Longview—Southbank Road Industrial Service)

CONVERSION TO BUSES

ALTERATION TO SERVICES ON ROUTES 10, 10B, 10C, 40 (and some alteration to Route Numbers)

Commencing SUNDAY, 22nd JUNE, 1952.

Journey time	43 minutes (10C)
Last day of operation	21 June 1952
Cars seen in service on last day	262, 272, 293, 885, 889, 890, 891, 897, 899, 902, 951, 988
Last car	891
Track abandoned	Liverpool Road reserved track beyond Pagemoss.

Longview Lane, last day view. *Leo Quinn/Online Transport Archive*

1952 | 141

22 June 1952: The day after the 10C conversion, a privately organised tour was held on 973. This had been fitted with a spare prefabricated end following an earlier collision with a Cabin car. The tour criss-crossed the shrinking network visiting Utting Avenue East, Kirkby, Muirhead Avenue, Pagemoss, Penny Lane and Garston. Despite missing out on the 10C, those on board were able to ride over routes 48 and 49. As the car turns from Harlech Street into Carisbrooke Road by Walton depot, Ted Gahan, point iron in hand, poses on the platform. *[Above] Photographer unknown*

During the scheduled visit to Edge Lane, the following withdrawn air-brake Standards were seen: 81, 91, 95, 97, 98, 128, 147, 328, 338, 343, 348, 440, 459. Also waiting to be dismantled were 502, 555 (ex-539), 777, 788, 820, 826, 834 and 836. Recalling better days, Green Lane regular 459 of 1932 turns onto Erskine Street with a peak hour 29 on 17 April 1951. Although reconditioned in 1937 with an EMB flexible axle truck and 60hp motors, its indicator displays were not modernised. *[Below] E A Gahan/Online Transport Archive*

23 June 1952: The few Monday to Friday peak hour journeys to and from Caird Street for Ogden employees now originate at Pagemoss. "The first time I did it I had a lively discussion with my driver who said I should show 10B. I was very unhappy about this travelling along West Derby Road and Green Lane. It seems this was the custom but whether it was mandated or just what the crews did I don't know." (Brian Cook) Officially, there was just one departure at 6.03pm but this may have been duplicated according to demand.

July 1952: "After a period of three weeks during which no EMB Standards are seen in service 723, 747 and 752 appear on peak hour duties." (JWG) A month earlier, 752 was photographed swallowing up a long queue at Commutation Row. *[Above] Peter Mitchell*

29 July 1952: 180 is hit in the rear by 249 on Muirhead Avenue East, the driver of 249, 23 year old James Cornish being injured. Wreckage quickly removed.

June-September 1952: During these four months, the last air-brake Standards were withdrawn. Based at Edge Lane these were: 723, 725, 727, 732, 747, 751, 752 and 754. This view of the upper deck of 732 was taken in June 1952. *[Below] Peter Mitchell*

5 August 1952: 751 was observed on Lime Street working a route 6. However, it is believed 747 was the last Reconditioned Standard to run in service although the exact date has not been recorded. On 1 October 1951, 747 is seen at Pagemoss. For a time during 1936/37, this car ran on Maley & Taunton swing-link bogies, later put under either 919 or 920. Note the steam waggon chugging towards Prescot. *[Page 144, bottom] Marcus Eavis/Online Transport Archive*

5 September 1952: Last day for Monday-Friday industrial services 42 (Penny Lane-Southbank Road) and 48 (Penny Lane-Gillmoss or Kirkby) although it is possible some unrecorded journeys may have taken place on Saturday 6 September to meet factory requirements.

142 | THE LEAVING OF LIVERPOOL

ROUTE PROFILE
42, 48

Edge Lane. *Peter Mitchell*

Edge Lane. *Peter Mitchell*

Mill Lane. *Peter Mitchell*

Muirhead Avenue. *Peter Mitchell*

Muirhead Avenue. *E A Gahan/Online Transport Archive*

Dwerryhouse Lane. *Peter Mitchell*

Tram Route 48, 49	KIRKBY, MUIRHEAD AVE. EAST —PENNY LANE	
Stage No.		Stage No.
57	Kirkby Estate	—
59	1½ Ormskirk Road	76
61	2½ 1½ Ainsworth Lane	74
63	4 2½ 1½ Radshaw Nook	72
65	4 4 2½ 1½ Gillmoss (Stonebridge L./Back Gillmoss L.)	70
67	5 4 4 2½ 1½ Lower Lane	68
69	5 5 4 4 2½ 1½ Oak Lane	66
71	6 5 5 4 4 2½ 1½ Lewisham Road	64
73	6 6 5 5 4 4 2½ 1½ Queens Drive	60
75	6 6 6 5 5 4 4 2½ 1½ West Derby Road	58
77	6 6 6 6 5 5 4 4 2½ 1½ Green Lane (Prescot Rd.)	56
79	7 6 6 6 6 5 5 4 4 2½ 1½ Binns Road	54
81	7 7 6 6 6 6 5 5 4 4 2½ 1½ Woolton Road	52
—	7 7 7 6 6 6 6 5 5 4 4 2½ 1½ Penny Lane	50

Journey time	17 minutes (42), 32 minutes (48)
Last day of operation	5 September 1952
Cars seen in service on last day	179, 203, 227, 286 (42); 87 (48)
Last car	Not known
Track abandoned	None

1952 | 143

"Cannot understand why the 49 is being replaced. Street track mostly in tip-top condition although parts are heavily-corrugated leading to lots of 'roaring' especially between the Clock Tower and Edge Lane Drive. Cars picking up and setting down at virtually every stop. Fun watching from the front seat as we squeal round Edge Lane roundabout and then more 'roaring' along St Oswald's Street. On this occasion, the conductor collects a staff controlling the single track section – then hops off to put it into the box on the other side. Maybe the electric signals weren't working. I don't remember. Later, I learnt some crews took a risk and went through without the staff when the lights were malfunctioning. Lots of jolting and jerking as my Baby Grand lurches through the worn point work in Old Swan. Quick glimpse into Green Lane – no trams, all out on the road. Nippy run along Green Lane and then a stop outside The Carlton. Driver and Inspector in conversation. Onto the reserved track for a fast run to the terminus – another last ride. Back into town on a 29." **(MJ)**

The all-street track 42 had four morning and one mid-day (1.12pm) departure from Penny Lane balanced by three from Southbank Road of which one was at 12.35pm and two in the evening. When introduced in 1932, the 48 was on all day service from Woolton but, after 1934, it became peak hours only. Latterly, there were four morning departures from Penny Lane of which the first went to Kirkby with the others terminating on the sidings at Gillmoss. In the evening, a single departure from Kirkby was complemented by two from Gillmoss.

6 September 1952: Last day for outer-suburban route 49 linking Penny Lane to Muirhead Avenue East via Church Road, Mill Lane, Edge Lane Drive, St Oswald's Street, Green Lane and Muirhead Avenue. With few through passengers, except at peak times, income was mostly derived from short-distance fares with many transferring onto other routes. Most of the line served different residential neighbourhoods with frequencies varying between 10 and 15 minutes except on Sunday mornings when it dropped to 30 minutes. Latterly, most duties were covered by Baby Grands although the occasional Liner did appear and, until recently, Reconditioned Standards at rush hours. Cars working to and from Edge Lane depot sometimes showed 42 or 48 on their screens instead of 49. Known as 'Abbey Specials', late evening cars would wait outside the giant Abbey cinema at Wavertree ready to take people home.

September 1952: SP1, formerly 30, was the first of four handbrake Standards to be used as snowploughs was assigned to Green Lane. Next came SP2 (646), SP3 (684) and SP4 (703) each being allocated to a surviving depot. They were painted in overall green, with their indicators plated over. When on active duty they were sometimes accompanied by a work car, a rubber-tyred salt trailer or even a Baby Grand, its platforms laden with bags of salt. Corrugation remained a noisy irritant. To alleviate the problem especially on residential streets, the four remaining 'scrubbers' made frequent nocturnal expeditions once normal service had ended. The first to be withdrawn, CE&S 273 is seen at Edge Lane in the summer of 1951. Since 1949, it had carried the cut-down body of Priestly Standard 636. *[Below] R W A Jones/Online Transport Archive*

June-September 1952: Following cars withdrawn from Walton depot: 772, 775, 799, 815, 828, 852. In mid-June 1952, 772 heads back to the depot via Queen's Road. *[Above] Peter Mitchell*

7 September 1952:

Green Lane allocation	157, 168, 169, 180, 188, 868, 875, 885-887, 889-891, 893, 897, 899, 901, 902, 904, 906, 907, 909, 911, 914, 916, 944, 946, 947, 950, 951, 961, 974, 985, 986, 988, 990

1952 | 145

ROUTE PROFILE
49

South Boundary Road, Kirkby. *D A Thompson*

Green Lane. *Peter Mitchell*

Muirhead Avenue East. *E A Gahan*

St Oswald's Street. *Peter Mitchell*

TRAM ROUTE
PENNY LANE 49 MUIRHEAD AV. EAST
INDUSTRIAL 42, 48 SERVICES

Journey time	24 minutes
Last day of operation	6 September 1952
Cars seen in service on last day	168, 205, 214, 226, 232, 263, 265, 271, 272, 296
Last car	263
Track abandoned	None. The section between Penny Lane and Edge Lane Drive was retained so Garston based cars could access the Works. Much of this street track had only recently been relaid and was in excellent condition. St Oswald's Street continued to be used by industrial services or by cars working to and from Edge Lane depot.

146 | THE LEAVING OF LIVERPOOL

Mill Lane. *Peter Mitchell*

Wavertree Clock Tower. *R S J Wiseman/National Tramway Museum*

Church Road. *Peter Mitchell*

1952 | 147

6 October 1952: From this date, the 19A was extended from Lower Lane to the city boundary at Gillmoss (Croxteth Brook) and peak hour cars extended to the loop serving Napier's factory given the number 19B which could also be used for short journeys on the 19/19A. *[Right, top] J B C McCann/Online Transport Archive*

Since the end of the war, traffic on the long semi-rural section from Lower Lane to Kirkby had declined dramatically. Sometimes hardly any passengers were carried off-peak whereas during rush hours it was three bells (for full up) and chain-on. The half-mile extension to 5 Gate did see some through cars but sometimes it only warranted a single car shuttling back and forth. In September 1951, 836 and 276 were both at 5 Gate terminus. Massive house building at Kirkby would bring much-needed traffic to this part of the system. *[Right, centre] J B C McCann/Online Transport Archive; N N Forbes/National Tramway Museum*

1 November 1952: Officially 56 cars scrapped during the year leaving 292 to work the remaining routes. Local enthusiasts believed the Corporation was again failing to maintain the fleet properly. Jack Gahan listed faults with individual vehicles which, in some cases, he forwarded to Hatton Garden. Here are some examples:

298	Both windscreens will not close. Indicator at one end will not turn. One standee strap missing. One door will not stay closed. Three seats on top deck without grabs or metal edging. Hook for trolleyrope missing one end.
788	Broken windows 'B' end Cabin – 25 March 1952.
820	Three lower deck windows jammed open. Rain comes in and resulting overlap very dirty. Badly dropped ends. Loose body with bulkheads coming away from sides. Interior very dilapidated. Bells not working – 7 April 1952.
164	Rebuilt body after being out of service four years but sent out in May 1952 with loose and rusty springs on trucks.
287	Although repainted recently, this car received no attention to its very loose bodywork. Outward appearance quite excellent but why was the body not attended to when in for its repaint?
185	Dreadfully dilapidated condition, badly sagging at the 'A' end. Trucks in bad shape and tyres extremely low. Noted in service 18 May 1952 with loose panel on lower deck. Track brake shoes missing.
239	Drastically defective bodywork, and although having steel strengtheners, leans about and creaks whilst travelling. The strengtheners are meant to hold the body firm but move about with the rest of it. The interior is in a dilapidated state. (This car later caught fire on Prescot Road on 11 June 1952).
901	Fitted with reconditioned seats from old type cars in November 1952 making the ride very uncomfortable. Also applies to 870.

946	Out after overhaul 26 September 1952. Dumped in Green Lane the same day and stayed there until 29 September. Out next day but making grating noises. Overhauled again during April-May 1953 but sent out with untouched trucks, the springs being rusty and loose

18 November 1952:

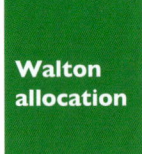

Walton allocation	151-156, 158, 161, 164-167, 170, 172, 175, 178, 181, 182, 185-188, 770, 774, 778, 779, 782-785, 789-798, 800, 802-804, 806, 807, 811-813, 816-818, 820, 823, 826, 827, 829, 830, 833, 835, 837, 839-841, 843-849, 851, 853, 855-860, 863-867, 877, 879, 880, 900, 905, 945, 948, 952-954, 956, 963, 965, 969-971, 973, 975, 977, 978, 981, 983, 984, 992.

Some of Walton cars were occasionally loaned to Garston to cover for their Maley & Taunton Liners which were still being reconditioned. For example, 153 and 178 were on loan in mid-November.

13 December 1952: 851 and 800 collide on Walton Hall Avenue.

148 | THE LEAVING OF LIVERPOOL

CHAPTER 10 | 1953

This was probably the last year in which the conversion programme could have been halted or revised. It made no financial sense to convert the Garston Circle. Even the press questioned such an obvious waste of public money, publishing photographs of the single-decker railcars recently delivered to Leeds Corporation. George Eglin of the *Liverpool Evening Express* began to write a series of pro-tram articles during the next few years and also to take part in the annual LRTL tour of the system. Following representations from Aigburth ratepayers, the Traffic Commissioners expressed their 'concerns' on the 23 March but to no avail. Although facing hefty increases in fuel tax, the Council were determined to be rid of the trams. Statistics were released giving the impression the Circle was losing £700 a week. There may have been some truth in this as the 'troublesome' Maleys were costly to maintain. However, they were good enough to attract the attention of Glasgow Corporation who agreed to the purchase of 24 cars on 1 June at a cost of £500 each plus £80 delivery. It is not known if anyone from Glasgow actually inspected the cars prior to their purchase. This sudden sale undoubtedly sealed the fate of the 8/33, conversion window bills appearing in the cars almost immediately.

19 January 1953: Approximately 20% of the 292 passenger cars out of service: 23 in depots awaiting repair with 40 in the Works, including 169, 182, 186, 187, 238, 239, 254, 268, 278, 293, 297, 815 (withdrawn) and 958. "The worst cars in this lot are subsequently the ones that come out overhauled!" (JWG)

1 February 1953: Garston allocation: 160, 162, 183, 184, 869, 870-874, 881, 883, 884, 901, 903, 918-942, 949, 955, 962, 966-968, 972, 976, 979 and SP4. (John Horne/JBH)

10 February 1953: Priestly Bogie 774 withdrawn. Following the six-car pile-up on Walton Hall Avenue, this had been rebuilt with three section indicator boxes at either end. Broken up on 20 May 1953, it is seen 11 months earlier leaving Gillmoss sidings.
[Below] Peter Mitchell

150 | THE LEAVING OF LIVERPOOL

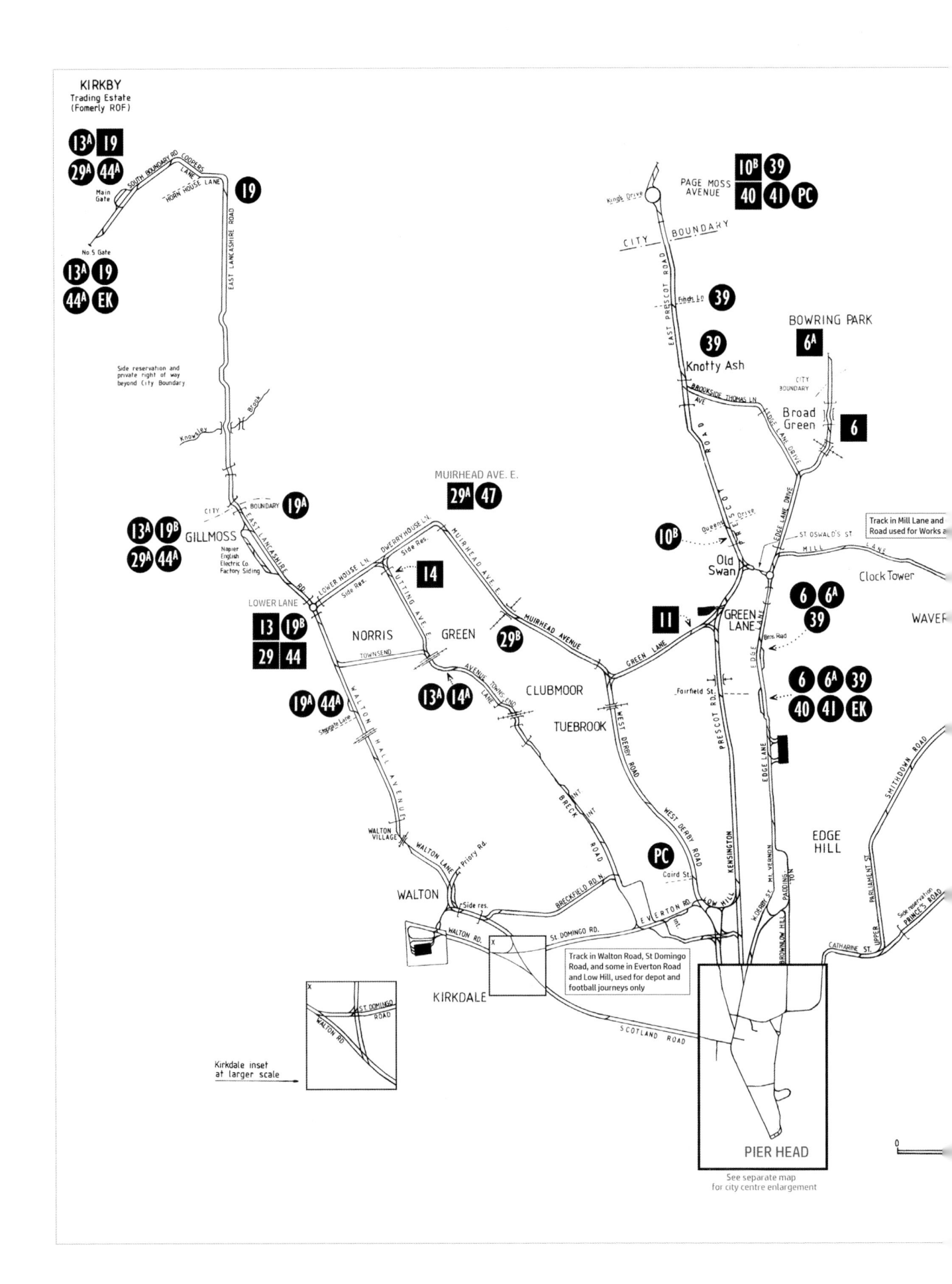

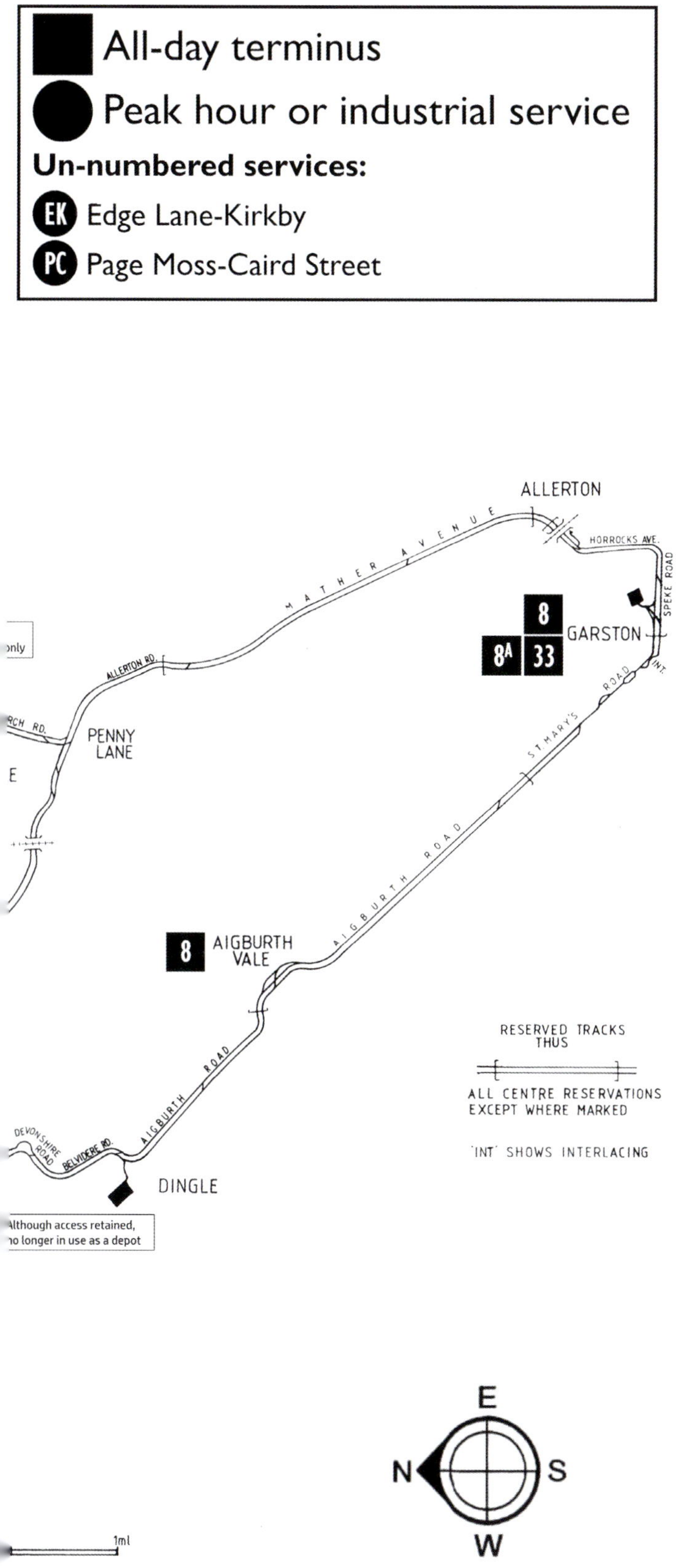

All-day terminus

Peak hour or industrial service

Un-numbered services:
- **EK** Edge Lane-Kirkby
- **PC** Page Moss-Caird Street

February 1953: 868, 873, 879, 920, 949 and 969 condemned as 'structurally unsound'. 868 was taken out of service in May 1952 although its trucks had only recently been overhauled. Despite its body being badly distorted, 920 had been 'patched up' following a collision. For its last few months, it served as Works shunter.

8 March 1953: SP4 noted ploughing from St Oswald's Street to Pagemoss at 10.50pm. On the return, its blade was reduced to simply digging up mud rather than dispersing snow. The crew stopped at Pilch Lane to tie up the blade with rope after which they proceeded slowly to St Oswald's Street arriving at 11.10pm.

12 March 1953: Route number 39 discontinued. This was used for all short workings of the 40. Although officially discontinued, some cars still showed 39 for several more months especially when returning to depot.

14 March 1953: 553, 566, 723, 800, 820, 826, 828, 851, 857 on the bank, of which some had been there for quite a while.

20 March 1953: To make way for the next influx of withdrawn cars, work started on clearing the dump with 820, 826 and former snowploughs 502, 553 and 566 moving to the scrap sidings at the rear of the Works where 502 still carries the legend 'Liverpool Corporation Tramways' on its rocker panel. Built at Lambeth Road it was in service from 1908 to 1938. *[Above] R W A Jones/Online Transport Archive*

The last Bellamy to be scrapped was 553 of 1911. After being withdrawn from service, it joined the snow plough fleet in December 1942. Like 502 it was on a Brill 21E truck and had two 40hp motors. *[Below] R W A Jones/Online Transport Archive*

1953 | 151

ROUTE PROFILE
39

Lord Street. *Alan B Cross*

Brownlow Hill. *E A Gahan*

152 | THE LEAVING OF LIVERPOOL

Edge Lane, passing entrance to the west side of the Works. *H B Priestley/National Tramway Museum*

Edge Lane Drive/Mill Lane. *R S J Wiseman/National Tramway Museum*

Edge Lane roundabout. *[Left] Peter Mitchell*

Keep your City tidy
—Please place used tickets in the receptacle provided, when leaving the vehicle.

SHELTERS have been erected at many stops—these are for your protection—please assist us to protect them from wilful damage

Journey time	Varied
Last day of operation	12 March 1953
Last car	Not known
Track abandoned	None

1953 | 153

28 March 1953: Grand National Day. For some unrecorded reason, routes 11, 13 and 29 diverted via London Road in lieu of Islington.

April 1953: Cars noted during a visit to Edge Lane:
Under repair: 203, 230, 263, 266, 264, 829, 846 and 967.
In the paint shop: 221, 245, 260, 911 and 917.
Awaiting scrap: 759, 761, 767, 769, 774, 800, 820, 826, 851 and 857.

The direct control English Electric bogie cars 759, 761, 767 and 769 were distinguished by their slightly sloping windscreens. 759 also had a modernised lower deck with ventilators in lieu of quarter lights whilst 769 had seats from scrapped Liners. As their controllers tended to overheat they mostly appeared at peak hours. By 27 May 1953, all were moved onto the bank awaiting scrap. 769 is seen on a peak hour working on Edge Lane Drive.
[Left] J W Gahan collection

17 May 1953: With the growth of high-density estates in the Kirkby area, all day service 19A was introduced. Operating every 10 minutes during most of the week, it linked the Pier Head to a newly created terminus at Southdene, an exposed spot, where a new crossover and Bundy time-clock were installed. This increased service put further pressure onto the electrical supply beyond Gillmoss. Concurrently, the 44 was extended from Lower Lane to the city boundary at Croxteth Brook with headways varying between 10 minutes in peak hours to 30 minutes on Sunday. Some 19As also terminated at Croxteth Brook and, in the peaks, a few 19As shuttled between Lower Lane and Kirkby. In March 1954, Cabin car 812 has arrived at Southdene. Note the use of the slipboard. From here, the car returned to Walton depot, which was usually shown as 'Spellow Lane' or 'Walton/Spellow Lane.' *[Below] J B C McCann/Online Transport Archive*

154 | THE LEAVING OF LIVERPOOL

22 May 1953: 868, the first Liner to be built, was broken up.

2 June 1953: Colourful bunting and street decorations for the Coronation of HM Queen Elizabeth II. Various celebratory events led to some interesting diversions. Overlooked by Rushworth and Dreaper, with its famous concert hall, 986, wrongly signed for 'Old Haymarket', is working an outbound 11 by way of a rarely used curve from Commutation Row into Islington. The car in the background is on a diverted 14. *[Above] J W Gahan*

6 June 1953: Known as the Garston Circle or Circular, the 8 and 33 departed from the Pier Head. The seven mile 8 reached Garston via Smithdown Road and Mather Avenue, the indicators being changed somewhere between Penny Lane and Garston to show 33 ready for the return to the Pier Head via Aigburth and Dingle whilst the seven-mile 33 reached Garston by way of Belvidere Road and Aigburth Road with indicators being changed to 8, possibly as soon as Princes Park Gates, ready for the return to the Pier Head via Smithdown Road. Cars travelling via Mather Avenue but terminating at Garston showed 8A. Early morning turns were often filled with dockers from the night shift at Garston Docks. There were now only occasional short workings to Clayton Square and Aigburth. In the past, there had been short workings on the 8/8A to Allerton Station (football), Smithdown Road (near the hospital) and to the junction with Lodge Lane where cars reversed in the evenings to pick up southbound audiences emerging from The Pavilion Theatre. The ticket overleaf was issued on the last day.

J B C McCann/Online Transport Archive

"After watching the lengthy coronation ceremony on a neighbour's tiny telly with about 40 others crammed into a small living room, I manage to 'escape' from home to ride the Garston Circle. By now I knew about the LRTL and had picked up on the grapevine that the 8 and 33 were threatened. Over the years, I had covered the Circle a number of times starting with a ride round on the 1 in about 1946. Since passing my Eleven Plus exam, my mother and I had to catch the 8 to reach an approved school outfitters on Smithdown Road. To me the Circle was a very fine tramway with miles of grass tracks all in good nick worked mostly by recently overhauled streamliners. Speeds – fantastic. Smooth, swift, stately running on full parallel on stable, solid, substantial track – Liverpool drivers were not shy of showing their paces. Rumours circulate about speeding trams chased by police cars and speeds in excess of 55mph. This was everything a modern tramway should be. I remember feeling genuine anger that it was all to be sacrificed on the altar of the all-conquering bus – couldn't people wake up to the folly before it was too late. After a 'tour' of Garston depot and a glimpse of SP1 it was back to town on board 936 this time via Aigburth on the 33. Whilst waiting to leave Garston, I 'borrowed for posterity' one of the colourful closure notices fixed to one of the upstairs windows (see left) – it is still a treasured possession. The next Saturday it was over. For me, this was the unkindest cut of all." **(MJ)**

LIVERPOOL CORPORATION PASSENGER TRANSPORT

TRAM ROUTES

GARSTON 8, 33 PIERHEAD

CONVERSION TO BUS OPERATION AND RENUMBERING 86 & 87

COMMENCING

SUNDAY, 7th JUNE, 1953

The Corporation had wanted to substitute buses on Sundays as from April 12 but the full service continued until early on Sunday 7 June 1953, although recently buses had covered for some duties on the 8A.

The last 8 (979) left the Pier Head at midnight followed by the last 33 (967) one minute later. On board 967 was Inspector Trevor who made a final ticket inspection before alighting at Dingle. As the tram approached Garston, a carnival atmosphere developed.

Outside the sheds, 979 was already waiting. "Inspector Stroud-Drinkwater was jostled into the centre of things for yet another round of 'Auld Lang Syne' before he cleared the way for the two trams to enter the depot, 967 first, followed by 979." (GWP) The

"There was dancing on the lower deck and a crowd, described as hundreds, swarmed around the track and the bumpers singing 'For he's a jolly good fellow' and 'Clang, clang, clang goes the trolley' accompanied by clashing of pan lids, tin trays and tambourines. Many people were moved to tears by the event. Dancing in the lower deck now turned into a carnival. The tram moved on its way, rolling down deserted St Mary's Road where a lone man doffed his hat as an amazed policeman surveyed the spectacle." **(Geoff Price/GWP)**

latter may well have been selected as it was the car featured on the colourful window bill. *[Below] Leo Quinn collection/Online Transport Archive*

The final shed allocation was recorded by John Horne. 160, 162, 183, 184, 869, 870, 871, 872, 873, 874, 878, 881, 883, 884, 901, 903, 918-942 (excluding 920), 949, 955, 962, 966, 967, 968, 972, 976, 979, SP1.

Leo Quinn/Online Transport Archive

Garston depot: In his book *By Tram to Garston*, Eric Vaughan remembers a few of the former staff: Wally Rankin (shed electrician), Marie Chandler (conductress, later Inspector), Percy Parsons, Sam Ellis, Roy Oultram (conductors), Tommy Pickering, Polly Drummond, George Humphries, Alf Humphries, Joe Bates, Matty Cowell, 'Grumbly Guts' and 'Hill Street Playboy' (drivers).

7-11 June 1953: During this period, the depot was quickly cleared, trams making a last journey via Horrocks Avenue, Mather Avenue, Church Road and Mill Lane. On arriving at Edge Lane Drive, the 24 Maleys were driven for storage inside the Works prior to leaving for Glasgow whilst 160, 162, 183, 870, 872, 874, 878, 883, 957, 962, 966, 967, 968, 976 and 979 continued to Walton depot by way of St Oswald's Street, Green Lane, Muirhead Avenue, Lower Lane, Walton Hall Avenue and Kirkdale Vale. 874 is seen by Wavertree Clock Tower. *[below] M J O'Connor/ National Tramway Museum*

11 June 1953: Last to leave was SP1, which is seen here at the empty Garston sheds. Surprisingly, no photos have come to light of this handbrake Standard making the last ever run along the magnificent Mather Avenue grass tracks as it made its way to Green Lane. *[Right] Leo Quinn/Online Transport Archive*

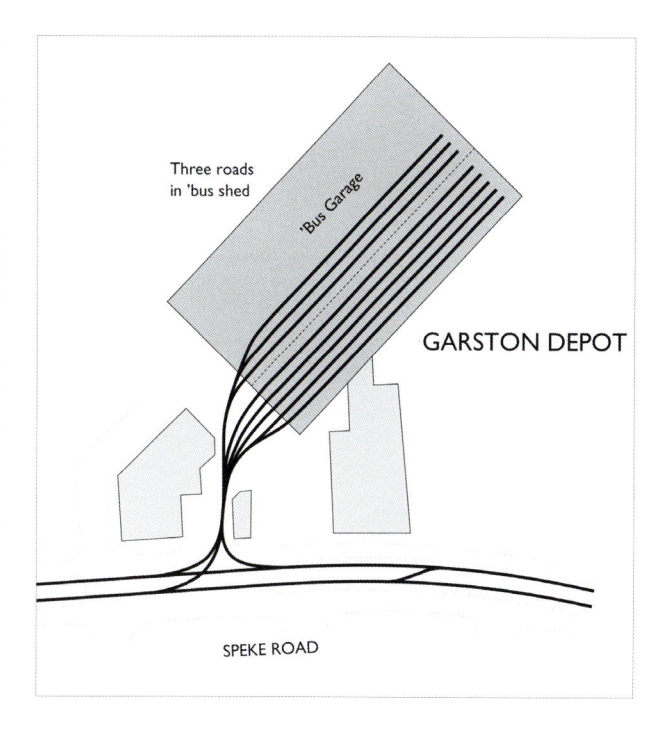

1953 | 157

ROUTE PROFILE
8, 8A, 33

Renshaw Street. *Peter Mitchell*

Upper Parliament Street. *Peter Mitchell*

Lodge Lane/Earle Road/Smithdown Road. *N N Forbes/National Tramway Museum*

Smithdown Road. *Peter Mitchell*

Mather Avenue. *K G Harvie*

Journey time	42 minutes (8), 38 minutes (33)
Last day of operation	6 June 1953
Cars seen in service on the last day	160, 162, 183, 870, 878, 918, 919, 921, 922, 926, 928, 931, 932, 933, 936, 937, 940, 957, 962, 966, 967, 968, 972, 976, 979
Last car	979 (8, J C Michaels (driver) and T Campbell (conductor)); 967 (33, A Trembath (driver) and S Hodgson (conductor))
Track abandoned	Renshaw Street, Leece Street, Catherine Street, Upper Parliament Street, Smithdown Road, Allerton Road, Horrocks Avenue, St Mary's Road, Aigburth Road, Belvidere Road, Princes Road. Also non-revenue trackage in Church Road and Mill Lane.

Mather Avenue/Burnham Road. *Peter Mitchell*

LIVERPOOL CORPORATION PASSENGER TRANSPORT
TRAM **8, 33** ROUTES
CONVERSION TO BUS OPERATION AND RENUMBERING **86 & 87**
COMMENCING: SUNDAY, 7th JUNE, 1953

158 | THE LEAVING OF LIVERPOOL

Horrocks Avenue. Extension opened July 1939, completing the Garston Circular. *J H Roberts/Online Transport Archive*

Speke Road, Garston. Note that the car still retained the two-line via destination. *Peter Mitchell*

ROUTE PROFILE
8, 8A, 33 (continued)

St Mary's Road, Garston. Note the interlaced track and the Coronation decorations. *R J S Wiseman/Online Transport Archive*

St Mary's Road, single track and loops section. *F N T Lloyd-Jones/Online Transport Archive*

Aigburth Road, Cressington Park. *Peter Mitchell*

Aigburth Road. *Peter Mitchell*

Aigburth Road, Dingle. *A S Clayton/Online Transport Archive*

Devonshire Road/Belvidere Road. *Peter Mitchell*

Princes Road. *Peter Mitchell*

Hardman Street. *Peter Mitchell*

160 | THE LEAVING OF LIVERPOOL

Many felt the heart had been sucked from the system. The Circle was a quality modern tramway with high-speed streamline cars providing a fast, quiet ride on newly-laid, segregated tracks. On the downside, off-peak loadings were light especially at the outer ends. The 14-mile Circular served more affluent suburbs with less compact housing. Also car ownership was on the increase. Determined to create a lasting memory local film-maker Alf Jacob took many precious minutes of ciné film. Subsequently, he recorded the remaining closures. His historic films survive and are shown to help raise funds for the Merseyside Tramway Preservation Society.

21 June 1953: Annual LRTL tour held on recently repainted 955; the 56 participants covering 52 miles in six hours. After a visit to Kirkby, the car reached Pagemoss by way of the 19, 29 and Green Lane in order to link up with the 10B which did not run on Sundays. From Pagemoss, it was into town via the 41 and 40 and then out to Utting Avenue East and Lower Lane before working across the city to Bowring Park for a final run into town. Here, 955 poses under the railway bridge at Clubmoor. The following year, the car was given HR2 trucks salvaged from Walton fire victim 965. Note the LCPT office in the background with crews waiting for changeovers. *[Above] R B Parr/National Tramway Museum*

July 1953: Depot allocations:

Edge Lane	201-208, 210-216, 218-224, 226, 227, 229-232, 235-255, 257, 258, 260-280, 283-289, 293, 296-299, 760, 762, 764-766 and SP3
Green Lane	157, 168, 169, 180, 885-887, 889-891, 893, 897, 899, 901, 902, 904, 906, 907, 909, 911, 914, 916, 986, 988, 990 and SP1
Walton	151-156, 158, 160-162, 164-167, 170, 172, 174, 175, 177-179, 181-188, 811-813, 817, 818, 823, 827, 829, 830, 833, 835, 837, 839-841, 843, 844, 847, 853, 856, 864-867, 869-872, 874-877, 878, 880, 881, 883, 884, 900, 903, 905, 917, 944-948, 950-958, 961-968, 970-979, 981, 983-985, 992. SP2 and SP4

Edge Lane continued to provide cars for routes 10B, 11, 29, 29A and 29B especially at peak hours. Cars were also occasionally loaned to Green Lane.

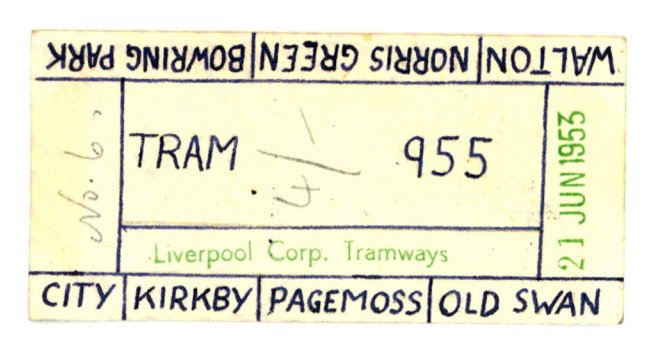

3 July 1953: Pagemoss to Caird Street workmen's service withdrawn. Run for the benefit of those employed in Ogden's Tobacco Factory this was the last unnumbered industrial service to appear in the timetable (Edge Lane-Kirkby did not). Advertised to operate 'according to factory requirements' it was 'liable to sudden alteration or cancellation'. Latterly, these duties were usually covered by cars from Edge Lane. In this view, 767 is seen on Green Lane on a working from Ogden's, showing 10B for Pagemoss. *[Above] J W Gahan*

Some 'Ogden's extras' on the 29 continued to reverse at Caird Street bound for the housing estates around Muirhead Avenue until April 1954. For a few days, 760, 762 and 764-766 in all-day use on the 6, 6A and 40.

4 July 1953: Last day of route 11 which basically provided additional capacity on the West Derby Road corridor. It was the last all street track route and the only all-day service to terminate in North John Street, cars arriving via Dale Street and departing via Church Street. Reflecting declining patronage, the basic headway was now every 30 minutes although supplemented by extras at peak hours and on Saturdays. Worked mainly from Green Lane, extras from Edge Lane appeared at peak times. Terminus usually shown as 'Green Lane' or 'Green Lane/Prescot Road'.

ROUTE PROFILE
11

Green Lane. *A D Packer*

Rocky Lane. *Peter Mitchell*

Outbound on one-way Brunswick Road, Gregson's Well. *Peter Mitchell*

162 | THE LEAVING OF LIVERPOOL

West Derby Road. Note the dangerous car driver! *A S Clayton/Online Transport Archive*

West Derby Road/Boaler Street. *A S Clayton/Online Transport Archive*

London Road. *A S Clayton/Online Transport Archive*

Lime Street. *J W Gahan collection*

LIVERPOOL CORPORATION PASSENGER TRANSPORT

TRAM ROUTE

GREEN LANE **11** NORTH JOHN ST.

CONVERSION TO BUS OPERATION

COMMENCING

SUNDAY, 5th JULY, 1953

TRAM **11** ROUTE

CONVERSION TO BUSES

COMMENCING

SUNDAY 5th JULY, 1953.

ROUTES & FARES REMAIN UNCHANGED

DETAILS IN HANDBILLS AVAILABLE AT ENQUIRY OFFICES

Journey time	25 minutes (inbound), 26 minutes (outbound)
Last day of operation	4 July 1953
Cars seen in service on the last day	257, 901, 904, 986
Last car	Either 901 or 904
Track abandoned	None

164 | THE LEAVING OF LIVERPOOL

6 July 1953: Cars inbound to North John Street from Muirhead Avenue East or Muirhead Avenue now showed 29A or 29B.

6 August 1953: Routes 19/44 diverted in both directions along the rarely used section of track on Spellow Lane. (JWG)

June to September 1953: Following withdrawn from Walton: 770, 778, 779, 782-785, 789-798, 800, 802-804, 807, 816, 845, 846, 848, 849, 855, 857-860 and 863. Surprisingly, 859 was loaned to Edge Lane during its final months. 770 and 778 known to have been on the dump by 5 July 1953. In the first view, 770 is inbound on Byrom Street and in the second 795 is on Shaw Street, both in June 1952. 795 was unusual. Following fire damage in 1935, it had been rebuilt as a Marks Bogie and mounted on EMB LW/1 trucks. *[Opposite] Peter Mitchell (both)*

When withdrawn, 770, 778 and 779 had lost their original side indicators and glass rain vents and had an additional metal block above the bumper to reduce structural damage in the event of a collision. 778 is at Lower Lane in June 1952. This car had a controller fitted with a brass plate saying 'no air track brake'. Note the open spread of countryside at this busy tramway junction. *[Right, top] Peter Mitchell*

12 September 1953: After being in store for three months, the first of the Maleys made the journey north to Scotland. Deliveries continued on a weekly basis until 24 March 1954, the trucks being forwarded in advance so they could be overhauled and re-gauged to 4ft 7¾in. Bodies, minus trolleypoles, were transported within a specially designed 'tram body trailer' acquired by Pickford's Heavy Haulage and fitted with internal padding to prevent damage to the bodywork. *[Below] JW Gahan/Ian McLean collection*

This was the order of departure: 927, 942, 934, 938, 935, 930, 931, 923, 928, 932, 940, 921, 922, 926, 937, 936, 918, 939, 925, 941, 924, 933, 929 and finally 919. One of these cars is shown on the traverser at Edge Lane being prepared for despatch. *[Overleaf, top left] JW Gahan/Ian McLean collection*

Early on each Saturday, the low loader travelled north via Salford, Bolton, Preston, Kendal, Carlisle and Beattock arriving on the Monday morning at Coplawhill Works in Glasgow. Here 927 is manoeuvred inside. *[Overleaf, centre] Leo Quinn collection/Online Transport Archive*

Staff at Coplawhill were "shocked by their poor mechanical and electrical condition". It was claimed an additional £1000 was spent on each car, much more than had been budgeted for.

Alterations included removal of the bumpers, fitting folding drivers' mirrors and bow collectors, painting the interiors and exteriors, completely rewiring the lighting and power supplies and revising the internal lighting. When completed, Coplawhill was confident each was fit for 20 years of service. Looking smart in its orange, green and cream livery, 1006 (formerly 942) is seen on Argyle Street shortly after entering service. *[Right, top] Ian McLean*

10 October 1953: Visit to the Works: 978 (new tyres), 299 and 231 (stripped down), 265 (bodywork and repaint), 298 and 224 (paint), 881 (varnish), 287 (replacement truck). (JBH)

31 October 1953: Another Works visit: 178 (varnish), 229 (truck and paint), 231 and 287 (awaiting overhaul), 235 and 907 (awaiting collision repair), 253 (minor body repair and paint). (JBH)

October 1953: Officially 219 serviceable cars.

7 December 1953: Route 44 extended to Southdene. This was the last windowbill advertising a route extension albeit over existing tracks.

166 | THE LEAVING OF LIVERPOOL

CHAPTER 11 | 1954

Despite knowing the battle to save the trams was lost, there was strong opposition to replacement of the 29 led by The Liverpool Transport Development Association (successor to the LTPA) headed by its outspoken Chairman Walter Purdy, Secretary Jack Gahan and Treasurer Roy Thomson. The LRTL published 6000 leaflets featuring a Liner on Glasgow route 29 under the title "Route 29 – but not in Liverpool". As usual, the Transport Committee was unimpressed. They stated yet again that the public preferred faster, more reliable buses. It was during this year, faced by acute staff shortages, that the Corporation once again employed conductresses (clippies), some of whom had served during the war.

15 February 1954: At 8.10am, 283 suffered a brake failure whilst descending Prescot Street. Driver Thomas Stevenson struggled with the handbrake as passengers start jumping off. Fortunately the car derailed thereby averting a serious tragedy as it would have struck another tram at speed. Having left the tracks, 283 mounted the pavement embedding itself in a traction pole outside the Falkland Arms pub on London Road fortunately without casualties. By 9.30am, it was re-railed and towed to Edge Lane. 283 was repaired and re-entered service with a new end believed to have been constructed in part from sections off 988. This was Liverpool's last major runaway although reduced maintenance may have led to less spectacular derailments, motor fires and brake failures. *[Below] Leo Quinn collection/Online Transport Archive*

January-February 1954: Snow ploughs occasionally on duty. Walton-based SP3 is at Kirkby. The points in and out of the loop have been cleared but the crossover, often used by drivers to reverse Baby Grands when controllers over-heated, is untouched.

[Above] Martin Jenkins collection/Online Transport Archive

Although Liners still suffered when snow turned to slush, 950 appears undeterred by a light covering as it crosses the junction of Muirhead Avenue and Queens Drive on 28 January 1954.

[Right, top] E A Gahan

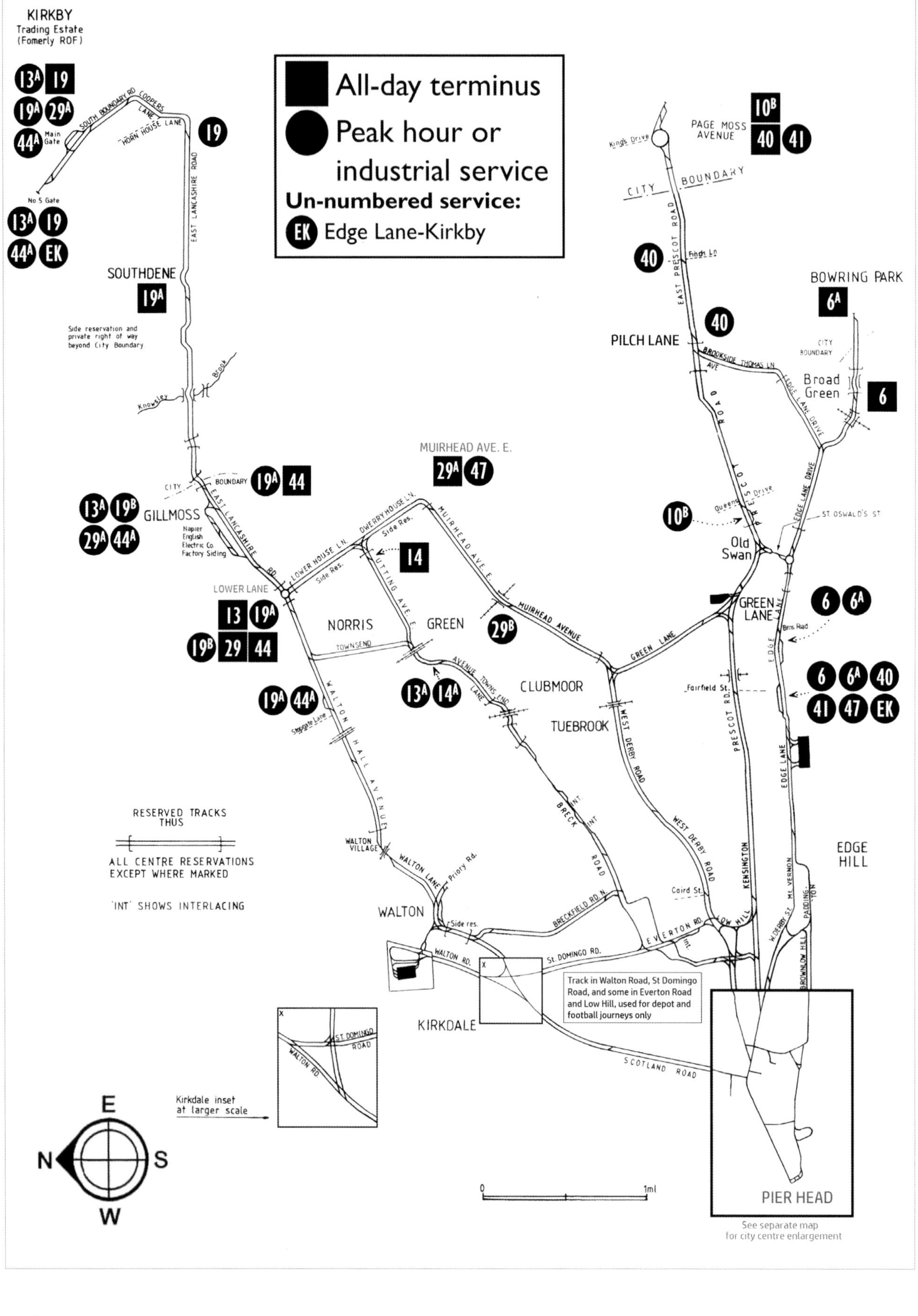

KIRKBY
Trading Estate
(Fomerly ROF)
All-day terminus
Peak hour or industrial service
Un-numbered service:
EK Edge Lane-Kirkby
PAGE MOSS AVENUE
SOUTHDENE
Side reservation and private right of way beyond City Boundary
BOWRING PARK
PILCH LANE
Broad Green
MUIRHEAD AVE. E.
GILLMOSS
Napier English Electric Co. Factory Siding
LOWER LANE
NORRIS
GREEN
CLUBMOOR
TUEBROOK
Old Swan
GREEN LANE
RESERVED TRACKS THUS
ALL CENTRE RESERVATIONS EXCEPT WHERE MARKED
'INT' SHOWS INTERLACING
WALTON
KIRKDALE
EDGE HILL
Track in Walton Road, St Domingo Road, and some in Everton Road and Low Hill, used for depot and football journeys only
Kirkdale inset at larger scale
PIER HEAD
See separate map for city centre enlargement
E
N
S
W
0
1ml

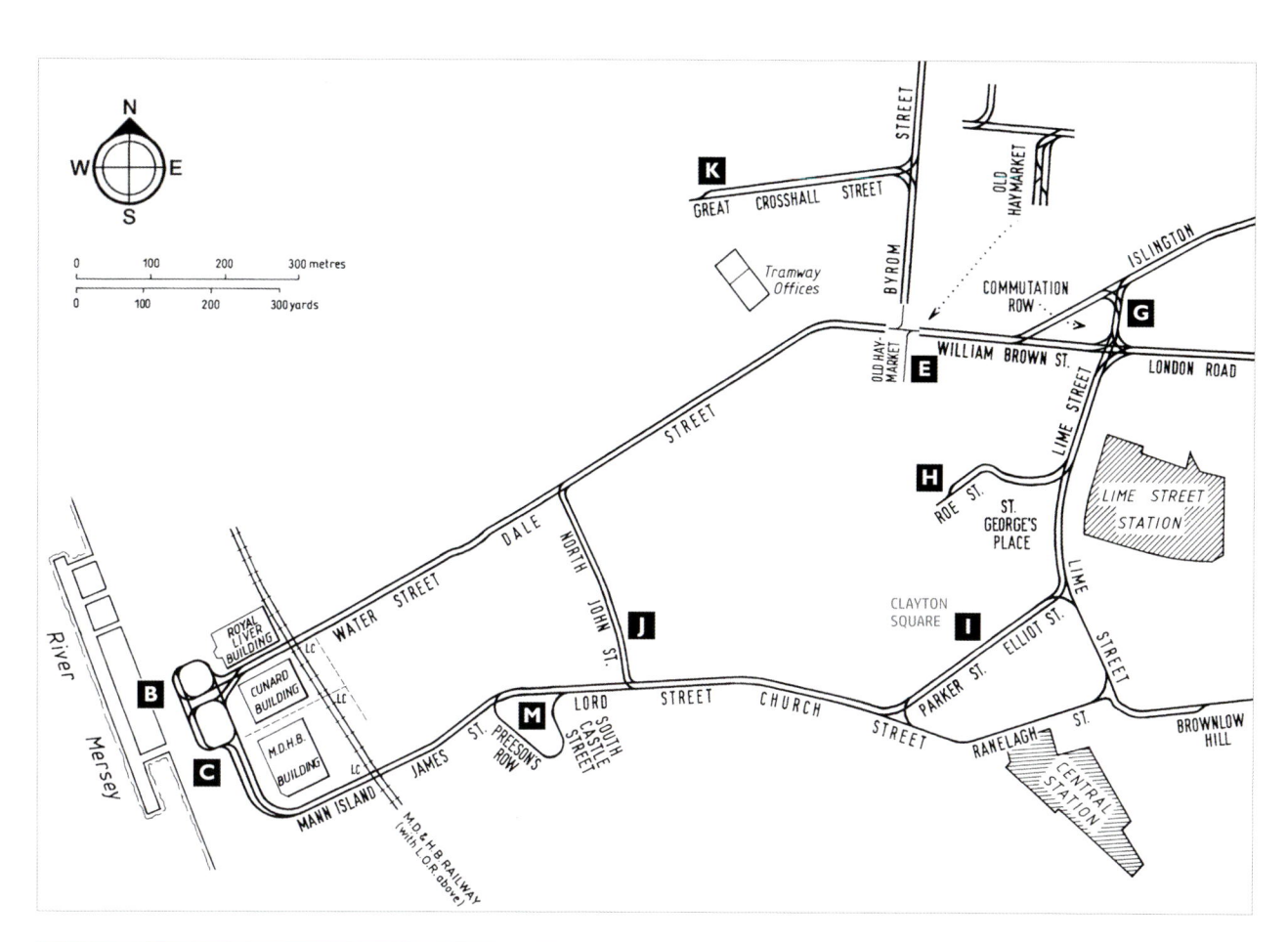

Key	Location	All day terminus	Terminal or start point for part-time routes
B	Pier Head (Centre Loop)	10B 29 29A	29B
C	Pier Head (South Loop)	6 6A 13 14 19 19A 40 44	13A 14A 19B 44A
E	Old Haymarket		13 13A 14 14A
G	Commutation Row *		10B
H	Roe Street *		6 6A
I	Clayton Square		6 6A 10B 14A 19A 19B 40
J	North John Street		13 13A 14 14A 29A 29B
K	Great Crosshall Street		44 44A
M	South Castle Street		One Sunday working only

* Cars for these points usually showed 'Lime Street' as the destination

28 February 1954: 60 participants on board 766 for a 45 mile Sunday excursion, at a cost to the LRTL of £6. Most of the day was cold and overcast with flurries of snow. Driver Harry Tindale of Edge Lane handled the 1931-built car with supreme confidence. From the pick-up point at Old Haymarket, he completed the nine-mile run to Kirkby via routes 13, 14, 29 and 19 in a startling 35 minutes! Then another high speed dash via the doomed 29 took 766 to the Pier Head from where the car followed the 44 towards Walton depot. The tour concluded with fast runs to Pagemoss via the 10B and finally Bowring Park before ending at Roe Street. Here, 766 poses on Village Street, part of the one-way system in Everton. *[Above] R B Parr/National Tramway Museum*

During the visit to Walton, one young member photographed 187 and 963 little knowing that within a few hours the latter would be involved in yet another depot fire. *[Above] A F Gahan/Online Transport Archive*

1 March 1954: 100 trams and 30 buses were in the depot when fire broke out on 983 at 6am. Kenneth Nicholls, the Shed Foreman, instructed staff reporting for duty to drive or tow vehicles out into the surrounding streets. Summoned on the direct phone line, the Fire Brigade was swiftly on the scene and the fire was soon contained although parts of the north bay roof collapsed. For several weeks, night watchmen kept an eye on 20 'evacuated' cars parked overnight with their doors locked on the siding on Walton Lane. As a result, the five English Electric bogie cars at Edge Lane earned a welcome reprieve. *[Below] Leo Quinn collection/ Online Transport Archive*

Ten cars, including 151 and 170, had broken windows and slight scorching due to the heat. These returned to service but 829, 840, 841, 844, 847, 965, 983 and 985 were towed to Edge Lane for scrap. Two of these casualties are seen in service in June 1952. 965 passes some of the pre-fabs erected along the East Lancashire Road whilst 983 is outbound on Breck Road with the first short section of reserved track visible in the background. *[Above] Peter Mitchell (both)*

The body of 963 was refurbished and placed on the EMB Lightweight trucks taken from under fire victim 829. In this view, it is on its original Heavyweight trucks. Other cars known to have been re-trucked during 1954/55: 188, 870 and 961 (EMB Lightweight), and 955 (EMB Heavyweight). *[Above] J B C McCann/Online Transport Archive*

985 suffered the most and is seen shortly after being towed to Edge Lane. *[Opposite, top left] Martin Jenkins/Online Transport Archive*

170 | THE LEAVING OF LIVERPOOL

March 1954: Glasgow Corporation purchased a further 22 Liners for £580 each – 15 on Lightweight and seven on Heavyweight trucks. It is not known why the latter were selected as Glasgow had no experience with this kind of truck. However, cars were probably chosen based on the state of their bodies, motors and tyres. Again it is not known whether any Glasgow official inspected the vehicles on offer. In this striking night time shot, one of those selected, 903 is seen at the Pier Head on 12 March 1954 alongside 964 which would become yet another fire victim in May 1955. *[Below] E N C Heywood*

2 April 1954: Liverpool had several unadvertised peak hour workings; for example, a few 29s ran from Edge Lane/Southbank Road to 5 Gate, Kirkby. The last to make this long trip was 273. To mark the occasion the crew signed their autographs for Tony Gahan. From 5 Gate, the car, its screen split between 29A/29B, worked to Clayton Square and then back to Edge Lane. On the same day, the 'Ogdens extras' from Caird Street were discontinued.

3 April 1954: Last day for service 47 which linked Southbank Road to Muirhead Avenue East via Old Swan and Green Lane.

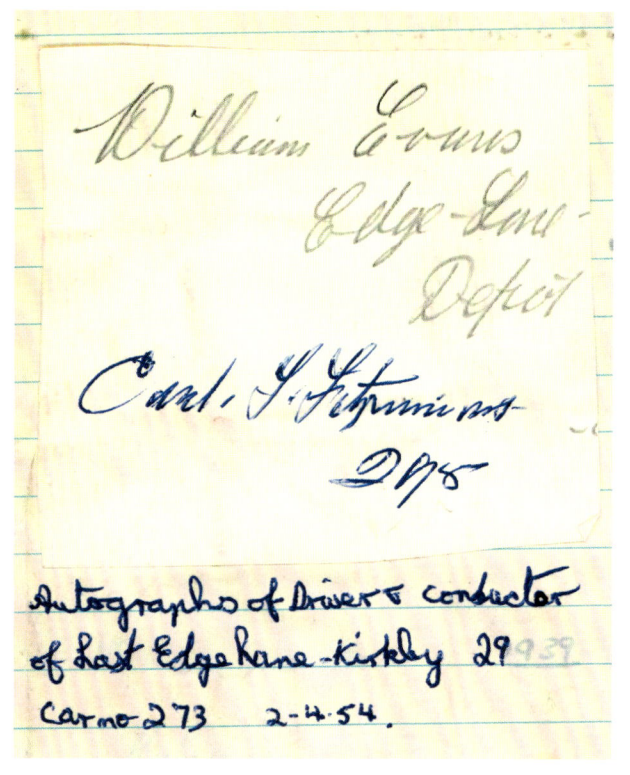

Introduced in 1932 and worked from Edge Lane depot, this had been an industrial service since February 1934 providing a useful link between residential areas in north-east Liverpool and the factories in and around Edge Lane. However, by 1954, there were only two advertised morning departures from Muirhead Avenue East and one from Muirhead Avenue Bridge. In the evenings, and on Saturday afternoons, two cars left Southbank Road for Muirhead Avenue East.

ROUTE PROFILE 47

LIVERPOOL CORPORATION PASSENGER TRANSPORT

TRAM 29, 29A ROUTES
EAST LANCASHIRE ROAD — MUIRHEAD AVENUE EAST — CITY
(AND ROUTE **47** INDUSTRIAL SERVICE)
CONVERSION TO BUS OPERATION

Journey time	20 minutes (approx)
Last day of operation	3 April 1954
Cars seen in service on the last day	201, 242, 263
Last car	242
Track abandoned	Joint with 29, 29A, 29B (see below)

Approaching Edge Lane roundabout. *Peter Mitchell*

The Carlton, junction of Green Lane and West Derby Road. *J B C McCann/Online Transport Archive*

Muirhead Avenue. *Peter Mitchell*

Muirhead Avenue/Queens Drive. *Peter Mitchell*

172 | THE LEAVING OF LIVERPOOL

The closure of the 6¼ mile route 29 marked the end of trams on West Derby Road and involved some five miles of track of which a high proportion had only recently been relaid. Starting from the Pier Head, Centre Loop and running via Dale Street and Islington, the 29s linked the city to Muirhead Avenue East (29A) and East Lancashire Road (29) as well as Gillmoss and Kirkby in rush hours. Most duties were worked from Green Lane but with some peak hour cars provided by Edge Lane. Inbound short workings ran to Green Lane and Edge Lane depots, Caird Street, Commutation Row, Old Haymarket and North John Street, the latter arriving via both Dale Street and Church Street. Some inbound and outbound short workings showed 29B. On Saturdays, additional 29s served the main shopping districts by operating inbound via London Road, Lime Street and Church Street before terminating in North John Street. A couple of photos exist of 29s bound for Church Street turning from Islington into Commutation Row prior to crossing into Lime Street. This may have been due to diversions. Headways varied from very frequent to every 30 minutes on Sunday mornings. Outbound 29s normally showed East Lancashire Road (sometimes abbreviated to East Lancashire Rd) whereas the 13s showed Lower Lane, Lower House Lane, Lowerhouse Lane or Norris Green/Lower Lane for exactly the same terminus. Occasionally, 29s also displayed one of these. 29As usually showed either Muirhead Ave East or Muirhead Av East.

Unlike the closure of the 8/33 very few people witnessed this closure. 891 left the Pier Head on the stroke of midnight and after a swift trip to Lower Lane it was in Green Lane depot shortly after 1.00am. Passing Carnegie Road bus garage, a few drivers sounded their horns in salute. In a quirk of fate, 891 (as No 1036) was also the last Liner to operate in Glasgow on their route 29!

Green Lane depot: For enthusiasts, Green Lane will always be associated with the 1947 fire and the restoration of 869 by the Merseyside Tramway Preservation Society (MTPS) between 1967 and 1979. John Horne recorded its final allocation by then housed on just the three eastern most tracks. Cars to be retained were sent to one of the two surviving depots with Walton receiving an influx of Baby Grands and some older drivers may have transferred to Edge Lane. Brian Cook, who was a conductor at Green Lane in 1952, recalls a couple of them: "Stan Higgins was a bus driver who also drove trams. He thought there was only one position on the controller – full parallel. I remember running in with the last 29 along the new track in Green Lane at about 50mph. A passenger sitting near the door said 'the man's mad'. One foggy morning, Stan was going from Muirhead Avenue to Lower Lane under the impression he was driving the first car. It turned out another had been diverted because of the fog and was parked by the roundabout. Stan hit it pretty hard and was relegated to conducting for three months". Brian also retains a culinary memory of Russell Warnes. "He lived in the lodge at Stanley Park and used to collect a snack from his house on the way to Pagemoss. It was with him that I first ate stuffed lamb's heart!"

4 April 1954: Some 15 Cabin and Marks Bogies listed for scrap were driven from Walton probably via Everton Road, Low Hill and Prescot Road for overnight stabling in Green Lane prior to proceeding to Edge Lane on the Monday morning.

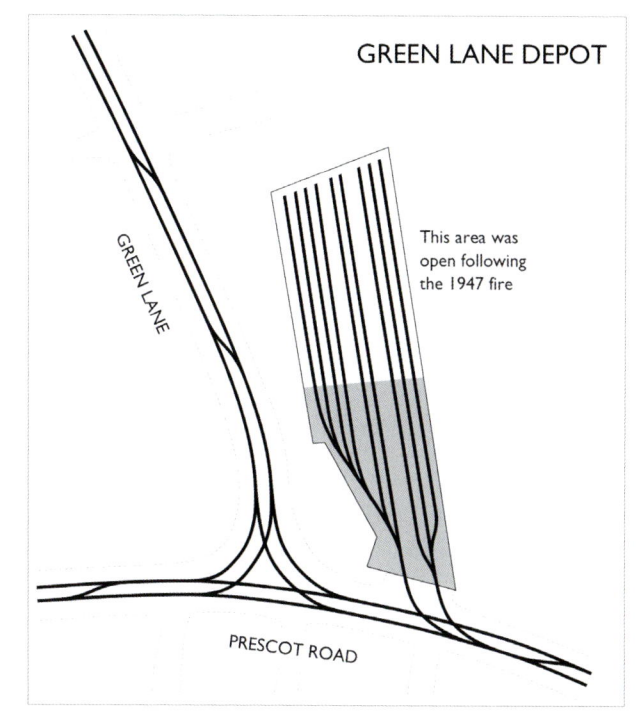

GREEN LANE DEPOT

This area was open following the 1947 fire

GREEN LANE

PRESCOT ROAD

GREEN LANE ROSTER. 3/4/54

157	891	914
168	893	986
169	897	988
180	899	990
875	901	
885	902	
886	904	
887	906	
889	907	
890	909	
	911	

ROUTE PROFILE
29, 29A, 29B

William Brown Street. *Peter Mitchell*

Islington Square. 216 outbound on Brunswick Road *(left)*, 269 inbound on Erskine Street. *J G Parkinson/Online Transport Archive*

West Derby Road. An 'Ogden's extra' reversing at Caird Street. *N N Forbes/National Tramway Museum*

West Derby Road/Caird Street. *Peter Mitchell*

West Derby Road, approaching Sheil Road/Belmont Road junction. *Peter Mitchell*

West Derby Road. *Peter Mitchell*

174 | THE LEAVING OF LIVERPOOL

Tuebrook. In this last day scene, note the off-centre trolley wire under the bridge and former tower wagon GKD320, now a mobile workshop and probably in use in connection with the removal of tram stops. *N N Forbes/National Tramway Museum*

Muirhead Avenue/West Derby Road. *J B C McCann/Online Transport Archive*

Muirhead Avenue. *J B C McCann/Online Transport Archive*

Muirhead Avenue bridge. *Peter Mitchell*

Dwerryhouse Lane. *J B C McCann/Online Transport Archive*

Lower House Lane. *Peter Mitchell*

Journey time	38 minutes (29), 33 minutes (29A)
Last day of operation	3 April 1954
Cars seen in service on the last day	201, 205, 207, 208, 210, 213, 216, 218, 242, 248, 269, 272, 273, 276, 296, 299, 885, 891, 899, 901, 902
Last car	891 (29), 248 (29A), 207 (29B)
Track abandoned	Pier Head (Centre Loop), West Derby Road, Rocky Lane, Green Lane, Muirhead Avenue, Muirhead Avenue East, Dwerryhouse Lane

1954 | 175

5 April 1954: Following closure of the Centre Loop (seen here on the last day), the 10B was moved to the South Loop and worked entirely from Edge Lane. *[Above] K P Lewis*

Only the South loop now remained. *[Opposite, top] Martin Jenkins collection/Online Transport Archive*

10 April 1954: Following cars on the bank: 817, 823, 827, 829, 833, 835, 839, 840, 844, 847, 853, 864-867, 965, 983, 985 and 988. Although officially sold for scrap in early 1952, Rail Grinder 273 was still on site. Still in sound structural and mechanical order, 839 works a peak hour duty to Southdene as late as 13 March.
[Opposite, bottom] J B C McCann/Online Transport Archive

27 April 1954: Whilst working the 5.27am from Broad Lane, smoke was seen rising from the floor of 174. Driver Grainger made for the wide part of Breck Road where conductor Caulfield saw everyone off. Within three minutes most of the body had been destroyed. The fire was blamed on an electrical fault.

12 May 1954: The first of the second batch of cars sold to Glasgow heads north. Following the same procedure as before, the trucks from 878 had departed six days earlier. The remaining 21 cars left in the following order with the final move on 5 November: 890, 869, 884, 885, 891, 899, 901, 902, 881, 883, 886, 871, 874, 877, 880, 893, 887, 903, 897, 904 and 875. Owing to a serious shortage of cars, some were kept in service until days before they left. Typically, 875 had just been overhauled and repainted! Although the 'Green Goddesses' were generally disliked in Glasgow, another approach was made in 1955 but, this time, no cars were available. *[Bottom left] Martin Jenkins collection/Online Transport Archive*

June 1954: For too long, 287 had been in appalling condition: its body and truck springs were dangerously loose and its windows moved from side to side. Although Jack Gahan wrote to three councillors during 1952/53 only its window winding apparatus was renewed! In July 1953, the car eventually entered the Works emerging 11 months later having been thoroughly reconditioned and fitted with girder type buffers from scrapped Liners, the only Baby Grand to have this feature. Seen at 5 Gate on 3 November 1956 it was the last tram to be completely rebuilt at Edge Lane.
[Above] Leo Quinn/Online Transport Archive

31 August 1954: Normally, works cars only ventured out at night but CE&S 283 enjoyed a short daytime spell reducing "roaring corrugation" on sections of the East Lancashire Road grass tracks. Some dished rail joints were also dealt with. Whilst grinding a given stretch, buses shuttled between each end of the curtailed tram service. When not in use, it parked up on the Napier siding with its trolley tied down. It is not known if it stayed there overnight or went to Walton depot or even back to its base at Edge Lane. *[Below] J Maher*

176 | THE LEAVING OF LIVERPOOL

1954 | 177

5 September 1954: Aware that the remaining Cabins and Marks Bogies would soon be withdrawn, the LRTL requested 818 for their second tour of the year but the Corporation refused, offering 766 instead. There was a sense of frustration as 818 had only recently been overhauled and repainted with the new-style fleet numbers. It is seen here on recently relaid track on Robson Street in June 1952. *[Above] Peter Mitchell*

Seizing his chance, Harry Tindale provided more high-speed thrills for the participants. The tour covered the remaining routes including the 10B which had no Sunday service. Harry was soon pacing the competing buses and easily overtaking them on the grass tracks. Departing from Old Haymarket, this was a classic Liverpool tour with the 50 or so enthusiasts enjoying runs to virtually all corners of the surviving system. Depot only sections such as Lowerhouse Lane and St Domingo Road were also visited as well as the sidings at Walton Lane, Napier's factory, Pagemoss, Southbank Road and Binns Road. The 56-mile tour ended at Roe Street. On this occasion, the sun was shining and Harry (right) posed proudly in his full uniform on the tram-only bridge over Knowsley Brook at Radshaw Nook. The second man is wearing the lighter summer dust coat which had just been introduced. 766 would remain in service for a further six months until withdrawal in early March 1955, after which its trucks were transferred to Liner 188. *[Right] Leo Quinn/Online Transport Archive*

For several years, senior *Liverpool Evening Express* journalist George Eglin had written features in support of the city's trams. Sensing a kindred spirit, the LRTL invited him to participate in their tours, and the resulting articles brought awareness of the interest in trams to a much wider audience.

"No 766 carried us for mile after mile and hour after hour. We defied all tramcar regulations. We smoked in the lower deck. We rang the bell for the driver to stop and start. We held up the tram while we got off to take photographs. We had a conductor aboard but we paid him no fares. And we even rode with the driver on his platform. Driver Harold Tindale was told to go as fast or as slow as we liked. We tried speed bursts and No 766 proved she could still skip along at 40mph. Oh Mr Councillor Bidston and your colleagues of the Transport Committee, how your ears must have burned. You were accused of not giving trams a fair chance in Liverpool. League members say you went a step in the right direction with the Green Goddesses and then got cold feet and refused to follow the modern trend in tramcars any farther. Even now, when you have declared against trams, they challenge you to give the Liverpool public the chance to see and try one of the single-deck super tramcars, such as are running in practically every city in the world outside Britain." **(George Eglin)**

As George described, there were many stops for photographs during the tour, including one on the section between Lower Lane and Utting Avenue East. Since the abandonment of the 29, this track was only used for occasional peak hour journeys.
[Above] Leo Quinn/Online Transport Archive

At 5 Gate, there was the obligatory group photograph, which features several individuals whose images or other contributions appear in this volume. On the left is Edward Piercey, one time LTPA Chairman, and next to him is J H (Bert) Roberts. Eighth from the left is Ted Gahan; his brother Jack can be seen behind, inside the car with an open-necked shirt, whilst youngest brother Tony is leaning out from the upper deck window. Shoulder-to-shoulder to the right of Ted is film-maker Alf Jacob, and a few places along with arms folded, is Bob Prescott. John Horne looks out from the seats by the second window in the lower saloon and in line with him, very slightly to the right, is the tall figure of Don Littler. Fourth from right is Harry Haddrill, whilst next to him (in the patterned jumper) is Leo Quinn. *[below] Leo Quinn collection/Online Transport Archive*

October 1954: Disgruntled local residents took matters into their own hands and built a crude shelter at Ribblers Lane on one of the most exposed spots on the East Lancashire Road. People in the new estates resented traipsing across muddy paths to reach this isolated stop especially after dark or on wet, windy days. Eventually, the Corporation relented and this concrete shelter was provided. *[Above] Phil Tatt/Online Transport Archive*

Christmas 1954: Realising it could not raise the funds to provide protective sheeting for Bellamy 558, the LRTL Museum Committee made the difficult decision to scrap the car. Its top deck was removed and the body transported to scrap merchants, Maden & McKee. One DB1 controller was salvaged and is now lodged with the MTPS. Other parts such as the dash and destination boxes may have been removed for preservation but have never surfaced. The scrap value enabled Norman Forbes to recover some of the cost of the transportation to Kirkby. This was a sad end to an early preservation attempt. If only someone could have put little 429 in their garden – what a Christmas present that would have been!

31 December 1954: There were now 172 passenger cars, four snowploughs, three works cars and two depots. A total of 18 cars were scrapped during the year.

CHAPTER 12 | 1955

The LTDA issued their first newsletter in January. Following pressure from the association, journey time on the 10B had been reduced from 37 to 34 minutes, which was still six minutes slower than the parallel buses. The association campaigned vigorously against closure of the 10B, the last along the Prescot Road corridor, issuing another attractive, well-argued leaflet. However, the newly appointed Chairman of the Transport Committee, Alderman Gregson, was unmoved, stating that the Committee remained firmly committed to its tram-scrapping policy.

19 February 1955: Roe Street closed as peak hour queue point for the 6/6A. Cars seen on the last day: 241, 247, 252, 278.

In the first Roe Street view, 751 and 213 are overlooked by St George's Hall and the former North Western Hotel on Lime Street as they await passengers in the autumn of 1951. Note the former horse car track leading into Queen Square. Sadly, this relic was lost when much of this historic market area was redeveloped. In the second, 254 picks its way cautiously through the fossilised junction with St John's Lane. By now all other wiring had been removed. *[Right] Phil Tatt/Online Transport Archive; Peter Mitchell*

20 February 1955: Supervised by Inspector Gordon Jones and crewed by a driver and two conductors SP4 was on points clearing duty. Towing an old rubber-tyred salt trailer, it made its way from Edge Lane along St Oswald's Street through Old Swan and out to Pilch Lane where it reversed in order to return to the Works via route 40. As recorded in his memorable book, *Edge Lane Roundabout*, Brian Martin was able to nip out of his house to take this remarkable view. He had been forewarned by Gordon Jones! *[Page 183] Brian Martin*

21 February 1955: Peak hour extras on the 6/6A now depart from Old Haymarket.

4 March 1955: Last day for another unadvertised service. To provide seats for those boarding east of Edge Lane roundabout, a handful of extras bound for Pagemoss loaded in St Oswald's Street on Monday to Friday evenings. To reach this boarding point, cars went from Edge Lane depot to Pagemoss via the 40 then via the 10B to Old Swan in order to access St Oswald's Street. In the past, they may have used the third track at Old Swan or the crossover at Finch Lane to reverse.

Remaining English Electric bogie cars bow out. 766 worked a peak hour 10B from Commutation Row to Pagemoss but it is not known if the others appeared during this Friday evening peak. Also there is no record of the final trips made by the surviving Cabins and Marks Bogies.

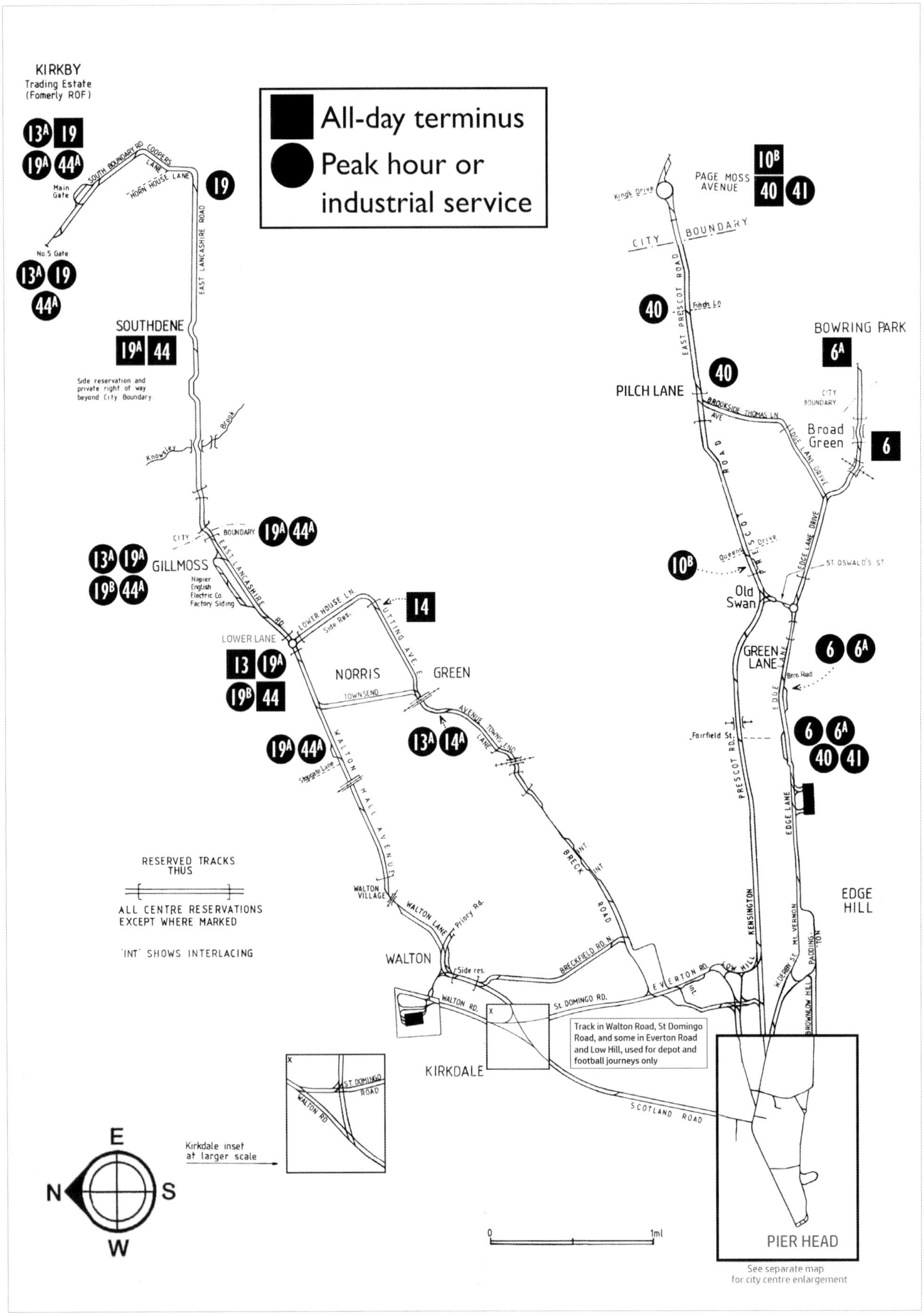
All-day terminus
Peak hour or industrial service
KIRKBY
Trading Estate
(Fomerly ROF)
13A 19
19A 44A
Main Gate
No. 5 Gate
13A 19
44A
SOUTH BOUNDARY RD
COOPERS LANE
HORN HOUSE LANE
19
EAST LANCASHIRE ROAD
SOUTHDENE
19A 44
Side reservation and private right of way beyond City Boundary
Knowsley Brook
CITY BOUNDARY
19A 44A
13A 19A
GILLMOSS
19B 44A
Napier English Electric Co. Factory Siding
EAST LANCASHIRE RD
LOWER HOUSE LN.
Side Res.
UTTING AVE. E
14
LOWER LANE
13 19A
19B 44
NORRIS GREEN
TOWNSEND
TOWNSEND AVENUE
LANE
13A 14A
19A 44A
WALTON HALL AVENUE
Stopgate Lane
BRECK ROAD
INT
RESERVED TRACKS THUS
ALL CENTRE RESERVATIONS EXCEPT WHERE MARKED
'INT' SHOWS INTERLACING
WALTON VILLAGE
WALTON LANE
Priory Rd.
WALTON
Side res.
BRECKFIELD RD. N.
WALTON RD.
St. DOMINGO RD.
EVERTON RD.
Int.
Track in Walton Road, St Domingo Road, and some in Everton Road and Low Hill, used for depot and football journeys only
KIRKDALE
SCOTLAND ROAD
ST. DOMINGO ROAD
WALTON RD
Kirkdale inset at larger scale
E
N
S
W
0
1ml
PAGE MOSS AVENUE
10B
40 41
Kings Drive
CITY BOUNDARY
EAST PRESCOT ROAD
40
Finch Ln
40
PILCH LANE
BROOKSIDE AVE
THOMAS LN
BOWRING PARK
6A
CITY BOUNDARY
Broad Green
6
EDGE LANE DRIVE
PRESCOT ROAD
Queens Drive
10B
ST. OSWALD'S ST
Old Swan
GREEN LANE
EDGE LANE
6 6A
Fairfield St.
6 6A
40 41
PRESCOT RD.
EDGE LANE
EDGE HILL
KENSINGTON
LOW HILL
W. DERBY ST.
MT. VERNON
PADDINGTON
BROWNLOW HILL
PIER HEAD
See separate map for city centre enlargement

"760, 762, 764, 765, 766, 811, 812, 813, 818, 837, 841, 843 and 856 are much sought after by enthusiasts and 'last rides' remain in the memory. During the previous two weeks I was lucky to enjoy 812 all the way to Southdene on a 44A and 841 on a peak hour 13A through to Kirkby. Both in good condition and ride well especially when full of passengers although too many of the wind-down windows are stuck, letting in water. They have a heavier, more stately feel than a Liner offering a steadier but slower ride. Great thrill – conductor on 841 allows me to stand on the back platform whilst he collects fares on the fully laden car. I feel quite important keeping the chain on and giving three rings on the bell as we approach crowded stops along the East Lancashire Road and sail past, trolleyrope billowing out behind. By Grant Gardens we are empty.

"From here the car rolls along the peak hour only track to the depot. I stand up front talking to the driver. On arrival, he takes me in to 'have a gander but mind dem der pits, lar'. I record seeing 811, 837 and 856. I also remember two rides on English Electrics one from Commutation Row and one from Pier Head. I had to make my way over from school and be in town before 5pm. I don't recall these cars appearing on Saturdays at least latterly. Historically, these were our last traditional trams and the last Priestlys still in service. A highlight – Harry Tindale with 764. We fairly hurtle back from Pagemoss. Cold, frosty, sprinkling of snow on the tracks – sparks and flashes from the overhead lighting the sky. Harry is certainly having a final fling. Up St Oswald's Street and round the roundabout and helter skelter down to the depot. On arrival, Harry smiles. 'Enjoy that?' A treasured memory." **(MJ)**

Don Littler

"On the last day of the 10B it was cold and a smattering of snow still remains. Determined to soak up as many memories as possible I join forces with Brian and Jeff Martin, Tony Gahan and Don Littler. Don takes a photo of us on the upper deck of 961 at the Pier Head. Hopping on and off we try to record cars at different vantage points. From Pagemoss we ride to Edge Lane depot on 249 which works back as a 41. During the afternoon, we encounter other enthusiasts including Tony's older brother Jack and the two Oldfield brothers who are taking ciné film. My last ride is on 950 – 'the mobile ruin'. I am not allowed to stay out for the very end as it would mean getting home about 2am." **(MJ)**

Brian Martin would go on to capture the memories of so many of these sad occasions brilliantly in his evocative book *Edge Lane Roundabout*.

ROUTE PROFILE
41

Route 41	PAGE MOSS—SOUTHBANK ROAD
Stage No.	Stage No.
14 Page Moss	
18 1½ Finch Lane	
20 2½ 1½ Eaton Road	15
22 4 2½ 1½ Blackhorse Lane	13
24 4 4 2½ 1½ Church Road	11
— 5 4 4 2½ 1½ Southbank Road	9
	7

Edge Lane. *Peter Mitchell*

LIVERPOOL CORPORATION PASSENGER TRANSPORT
TRAM ROUTES 10B, 41 CONVERSION TO BUS OPERATION

Old Swan, St Oswald's Street. Note the classic Liverpool tram stop - red for compulsory, blue for request. *J B C McCann/Online Transport Archive*

Old Swan. Note the peak hour siding on the right. *Frank Oldfield/Online Transport Archive*

Knotty Ash. *J B C McCann/Online Transport Archive*

Journey time	18 minutes
Last day of operation	5 March 1955
Cars seen in service on the last day	249, 277
Last car	277, but 249 as the last 10B displayed 41 when returning to Edge Lane depot
Track abandoned	St Oswald's Street

184 | THE LEAVING OF LIVERPOOL

5 March 1955: Trams removed from the Prescot Road corridor, once the busiest section on the system. Some four route miles abandoned. During the final afternoon, 811, 812, 813, 818, 837, 841, 843 and 856 were driven from Walton to Edge Lane via St Domingo Road, Low Hill, Prescot Road and St Oswald's Street. 812 was the last Cabin and 856 the last Marks Bogie to make this trip.

The 41 from Southbank Road to Pagemoss via Old Swan was another industrial service. Unnumbered from 1936 to 1947, it transported workers from the council-owned housing estates in and around East Prescot Road to the labour-intensive factories flanking Edge Lane. Latterly, there were a couple of morning departures from Finch Lane plus two at mid-day from Pagemoss. These were balanced by one mid-day and six evening departures from Southbank Road. Officially, the 41 only ran at certain times on Mondays to Saturdays. In reality, cars returning to depot from the Pagemoss direction usually showed 41 instead of 10B if running in via St Oswald's Street.

Since June 1952, four cars maintained the off-peak 20 minute service on the 10B (Pier Head to Pagemoss via Old Swan) but, at peak times, up to 15 or 16 extras could appear. These were needed to supplement the overcrowded replacement buses. In the evenings, cars arrived at the main evening queue point on Commutation Row (shown as 'Lime Street' on screens) by way of Low Hill, Erskine Street and Islington. Other 10Bs short-worked from Fairfield Street, Low Hill, Clayton Square and North John Street. Heading east towards Pagemoss, cars also turned short at Old Swan, Finch Lane and Pilch Lane. Since April 1954, the route had been worked by Edge Lane depot. There was no Sunday service.

The last 10B left Pier Head at midnight. A small knot of enthusiasts and well-wishers were on hand at Pagemoss including Brian Cook, one-time conductor at Green Lane. He took this view shortly before 249 left for Edge Lane at 12.37am. Displaying the number 41 it became the last tram to tackle 'the ruts' (the appalling stretch of street track in Old Swan) before heading up St Oswald's Street and down Edge Lane. For decades, crews talked of working 'A Moss' or to going to 'The Moss'. *Brian Cook/Online Transport Archive*

At the time of the 10B conversion, the LTDA produced another four page handout condemning the closure. Two of the LTDA members, Frank and Ron Oldfield produced a short film *The Black Horse Will Always Remember*, the Black Horse being a well-known hostelry at Old Swan. This precious 9.5mm film survives.

ROUTE 10B

Trams to be abandoned!
THIS WILL MEAN
Inconvenience to you.
THE REMEDY IS IN YOUR
OWN HANDS.

PROTEST
TO YOUR CITY COUNCILLOR

7 March 1955: The unadvertised peak hour departures to Pagemoss which formerly loaded in St Oswald's Street now left from Edge Lane roundabout.

Many of the remaining Liners and Baby Grands had loose and leaking bodies, windows and doors that failed to close whilst the poorly maintained trucks clattered and banged over the increasingly neglected tracks. Some Baby Grands continued to be overhauled, varnished and painted and 25 Liners were selected for refurbishment. 151-154, 156, 158, 160, 165, 172, 182, 186, 188, 870, 889, 907, 911, 916, 945, 947, 948, 953, 971, 984, 990 and 992 had their bodies strengthened, defective ventilators removed or replaced, leaks sealed, seats, flooring, lampshades and wiring renewed and new tyres fitted.

The last to be rehabilitated was 870, the oldest in the fleet, which was repainted on 14 September 1955. Each Liner had an interesting history and 870 was no exception. In January 1947, it received the bogies from 972. Then, following a bad collision later that year, it was laid up until 1951 when it went through the Works and was substantially reconditioned. It emerged with sliding as opposed to wind-down windows, its rain shields and side indicators removed as well as nearly all its external vents and is seen in this condition on Upper Parliament Street in June 1952. Following another accident on 16 February 1954, it re-entered service on 7 May with EMB Lightweight trucks off 867, high speed motors and seats probably from a withdrawn Cabin car. In May 1956, it was the last tram to be involved in a fatality. *Peter Mitchell*

ROUTE PROFILE
10B

London Road. *J B C McCann/Online Transport Archive*

Kensington. *R J S Wiseman/National Tramway Museum*

Prescot Road, Fairfield Street crossover. *R J S Wiseman/National Tramway Museum*

Prescot Road. *M J O'Connor/National Tramway Museum*

Stanley station. *Peter Mitchell*

186 | THE LEAVING OF LIVERPOOL

Outside Green Lane depot, with cars being led in at the end of peak hour service in April 1954. *J B C McCann/Online Transport Archive*

Prescot Road, east of Old Swan. *Peter Mitchell*

Knotty Ash, near Alder Hey hospital. *Peter Mitchell*

East Prescot Road, near the City Boundary. Note the prefabs on the left. *Peter Mitchell*

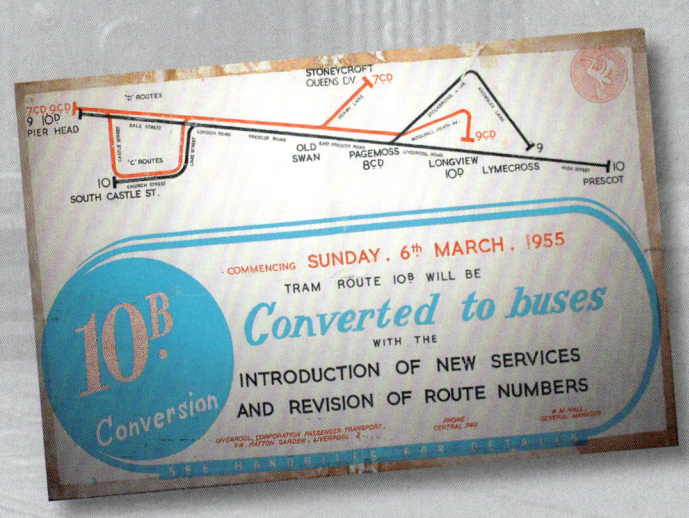

Journey time	38 minutes
Last day of operation	5 March 1955
Cars seen in service on the last day	188, 240, 246, 249, 257, 260, 262, 263, 267, 273, 277, 279, 906, 950, 961, 986, 992
Last car	249
Track abandoned	East Prescot Road (west of Knotty Ash), Prescot Road, Kensington; parts of Low Hill, Prescot Street. Commutation Row remained wired in case of emergencies until November 1955.

1955 | 187

25 May 1955: Whilst in service on Walton Lane, 964 burst into flames. The badly burnt body was towed to Edge Lane and placed on the dump. Here, the car negotiates the interlaced track on one of the narrow sections of Breck Road. *[Below] Peter Mitchell*

30 May 1955: Surprisingly, Brunswick Road and Erskine Street were reactivated when routes 13, 14 and 19 were diverted along Everton Road due to a religious procession. For years, lengthy parades on St Patrick's Day and by various Orange Lodges, caused major disruption. Routes were curtailed and diverted with connecting buses sometimes provided. Here 971 emerges from Everton Road into Grant Gardens with a diverted 14. It will make its way into town by way of the single track on Low Hill and then down Erskine Street. *[Right] F N T Lloyd-Jones/Online Transport Archive*

12 July 1955: Brunswick Road, Low Hill and Erskine Street again in use during Orange Lodge diversions. (Tony Gahan/AFG)

5 August 1955: Brunswick Road, Low Hill and Erskine Street used for the last time as a result of Orange Lodge activity. (AFG)

August 1955: Up to now, £900,000 had been spent on lifting track and resurfacing 30 miles of former street tramway. The cost for lifting a mile of sleeper track and removing fences and hedges was given as £4500 a mile.

31 August 1955: Death of R J Heathman, who had been deeply involved in the designs of the modern fleet.

188 | THE LEAVING OF LIVERPOOL

4 September 1955: Recently repainted and rehabilitated 165 was used for a five hour, 45-mile LRTL tour. Approximately 60 members took part with Harry Tindale again in control. Starting from Old Haymarket, all remaining sections of track were covered. R B Parr arranged this memorable photo opportunity on Edge Lane. With 165 occupying Southbank Road siding, service cars 169 and 911 were stopped and carefully positioned on the through tracks. The highlight was a fast, exhilarating run along Scotland Road with 165 leaving a trail of flying litter as it kicked up the chaff and dust. *[Above] Leo Quinn/Online Transport Archive*

September 1955: Car allocations:

Edge Lane	153, 154, 156, 164, 180, 188, 240, 244-254, 257, 258, 260-267, 271-273, 275, 278-280, 283-286, 288, 289, 293, 299, 872, 906, 907, 911, 916, 945, 950, 977, 986, 992
Walton	151, 152, 155, 158, 160-162, 165, 166, 169, 170, 172, 177, 178, 181-187, 201-208, 210-216, 218-224, 226, 227, 229-232, 235-239, 241-243, 255, 268, 270, 274, 276, 287, 296, 297, 870, 889, 905, 909, 917, 946-948, 951-953, 956-958, 962, 963, 967, 968, 970-974, 978, 979, 981, 984, 990

25 September 1955: 157, 168, 175, 179, 914, 961, 975 and 976 joined 811, 813, 837, 841 and 964 on the dump. Waiting their turn for the torch were 760 (the last of the class to be scrapped), 843 and 856 whilst laid up in the Works were 944, 955 and 966. Cars were now being withdrawn on the slightest pretext including low tyres, loose bodies, leaks and electrical faults.

4 October 1955: 944, 954, 961 and 975 now on the dump. 841 was the last Marks Bogie to be broken up and SP3 appears to have been withdrawn after use as work's shunter.

26 October 1955: Shorn of platforms and trucks, the lower saloon of 762 was installed as a bowling green pavilion at Newsham Park.

1 November 1955: Ready for the onset of winter, SP4 sent from Edge Lane to Walton during the height of the evening rush hour. Wrapped against the cold, the driver took the old handbrake car along route 40 to Clayton Square from where he headed north to Walton via the 19 as far as Everton where he followed the peak hour only track down Kirkdale Vale and along Walton Road.

5 November 1955: Replacement of routes 13/13A and 14/14A serving the Everton and Breck Road corridor. The six-mile 13 left the Pier Head and travelled to Lower Lane via Dale Street, Islington, Shaw Street, Breck Road, Townsend Lane, Townsend Avenue and Walton Hall Avenue. Much of the middle section was shared with the six-mile 14 which linked the Pier Head to Utting Avenue East by way of Church Street, London Road, Shaw Street, Townsend Lane and Utting Avenue East (sometimes Utting Ave East or Utting Av East on screens). Operation of the two routes was interlinked with most 13s arriving at Pier Head departing as 14s and vice-versa. Each served densely-knit communities and featured examples of one-way working in nearby streets, single and interlaced track as well as several miles of reservation mostly serving between-the-wars estates at Norris Green. Depot journeys to and from Walton, plus a variety of short workings, showed 13A/14A. On Monday-Friday evenings and Saturday lunch-times, the main queue point was Old Haymarket (seen above) although a few 14As started from Clayton Square. North John Street was used on Monday to Saturday mornings by a handful of inbound cars on both routes. Outbound 13/14s occasionally terminated at Cabbage Hall and Broad Lane and during Monday to Saturday peak periods there were industrial extras to Gillmoss and Kirkby.

Crews returning to Walton depot from East Lancashire Road had a choice of three routes: via the 19 as far as Everton Valley and onto Kirkdale Vale; via the 13 as far as Aubrey Street, then Everton Road as far Grant Gardens, reverse and then via St Domingo Road, Kirkdale Vale and Walton Road; or finally via the short surviving length of side reservation along Lower House Lane, then onto Utting Avenue East as far as Broad Lane from where drivers followed option 2. When heading for depot, cars usually showed 'Spellow Lane' or 'Walton/Spellow Lane' the other screens set to blank. Crews usually referred to the depot as 'Spellow' and those assigned to the 13s and 14s worked 'The Cabbage' or a 'Cabbage' referring to Cabbage Hall Inn, a local pub in the Anfield area. A view taken in June 1952 shows a car with a 'Cabbage Hall' slipboard in the driver's window.

North John Street was last used during the morning peak when 297 headed into Lord Street as a 14 and 219 into Dale Street as a 14A at 8.02. This important link between Lord Street and Dale Street was completed in 1938 and was served by several routes over the years and, until 1950, by Football and Races extras accessing Victoria Street. Here, a couple of diverted Baby Grands emerge onto Lord Street in 1954.
[Below] Martin Jenkins collection/Online Transport Archive

Last departures from Old Haymarket: 178 (14), 205 (13) and finally 272 (6A) which left for Bowring Park at 1.30pm. This view dates from March 1954. *[Top, left] J B C McCann/Online Transport Archive*

The last 13 was 182 which left Pier Head at 11.55pm whilst the last 14, 158, left at 12.02am. There was virtually no public interest and only a few committed enthusiasts were on board. Stage Two of the conversion programme was now complete. This night scene was taken shortly before the end. *[Opposite, centre right] J B C McCann/Online Transport Archive*

"This was a significant closure for the small group of enthusiasts who met regularly on Saturday mornings to digest the latest transport news in the area. Since late 1953, I had become part of the group. Our favoured meeting place was the end of platform 7 at Lime Street where Jack Gahan would regale us with all his news. He was our 'guru' and kept copious notes. Sometimes, there would also be 'inside information' from Ted Fowler who was employed in the works. He would sidle up to us and deliver his tit-bits as if they were state secrets. From there, we would snatch a quick drink (we usually took sandwiches to save money) before making for Old Haymarket to monitor the peak hour extras.

"Prior to joining the gang, I had often positioned myself at the junction of Lime Street, London Road and Commutation Row so I could see what was working all the many additional Saturday peak hour journeys. Sometimes, I would opt for a fast ride on an English Electric on a 6, 6A or 10B; on other occasions it would be a Cabin or Marks. In my notes I have recorded riding 779 on a 14A but now I have no memory of this although I do remember riding on 774, the rebuilt Priestly bogie car, on a 13A to Broad Lane. So it was that on 5 November we broke off from photographing the 13/14s so, with Jack Gahan, we could monitor the final departures from Old Haymarket. All seemed so permanent. Here was the usual Saturday mid-day crowds forming orderly queues with inspectors marshalling the 'extras' into their correct loading areas.

"When 272 left, it all came to an end. I spent the rest of the day with Brian and Jeff Martin, Ronnie Stephens, Tony Gahan and Don Littler. We had learned from photographers such as Henry Priestley, Richard Wiseman, 'Freddie' Lloyd-Jones and Norman Forbes that you had to walk a route and photograph all the strategic locations. However, the Breck Road services had so many fascinating features all of which cried out to be captured on film – steep inclines, one-way workings, single track, interlaced track and sections of reservation. Also, as on other routes in the east of the city, the grass tracks were often impeded by narrow railway bridges where the tracks had to return to the highway in order to get through, the railways having simply refused to widen where necessary. We took what we could but film was expensive and had to be rationed so we tried not to duplicate each other. We thrilled at the speeds along Utting Avenue East and Townsend Avenue and felt annoyed the trams were coming off.

"However, we were also aware that many Liners were especially grubby and neglected with ripped seats, jammed windows, loose bodies, peeling paint and water dripping in at various places. BUT we still knew they were superior to buses which relied on imported fuel. After a series of final runs, I had to make my way home leaving the Pier Head shortly after 9pm. Another two routes had gone and we would no longer congregate at Old Haymarket." (MJ)

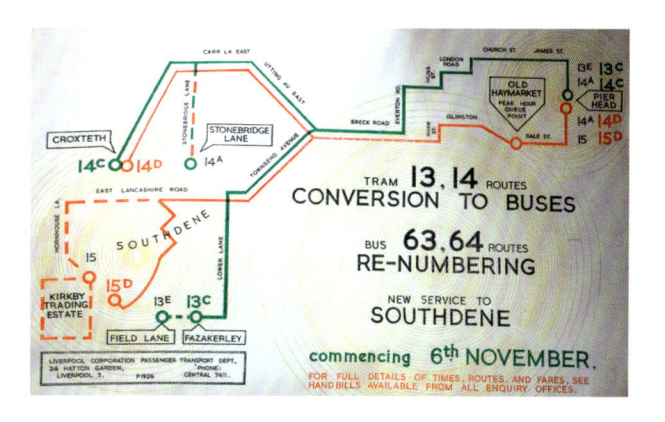

An internal memorandum has survived entitled "Tram-Bus Conversion R.13/14. Operative Sunday 6 November 1955. Revised allocation and traffic details". Buses and trams listed for scrap are shown as 'To Pool'. "Movement. All by arrangement with D/ Inspectors. Tramcars for Pool to be ferried to Edge Lane Works on Saturday and Sunday Nov 5-6, & Tram cars for transfer to Edge Lane depot. Foreman on duty at Walton and Edge Lane responsible for reception and despatch. Indicators. All screens, Lamps and Platform Equipment to be removed from 'Pool' cars and taken into Depot Stocks. In the event of cars for transfer from Walton not having comprehensive screens. Engineer to notify Edge Lane Depot as soon as possible."

Walton's allocation was now reduced to 64 cars for a peak demand of 52, all of which were housed on the five middle tracks. 201-208, 210-212, 956, 973 and 974 were transferred to Edge Lane which needed 61 cars for its peak demand of 51. Similar instructions were probably issued for each conversion but no others appear to have survived. Now, they would be invaluable.

ROUTE PROFILE
13, 13A, 14, 14A

James Street, taken from Liverpool Overhead Railway James Street station with James Street Mersey Railway station in backgound. *W J Wyse/LRTA London Area/Online Transport Archive*

Peak hour 14A turning from Church Street to Parker Street. *J B C McCann/Online Transport Archive*

Islington. *Peter Mitchell*

Eastbourne Street/Shaw Street. *Leo Quinn/Online Transport Archive*

As part of the Everton one-way workings, this stretch of Fitzclarence Street was used by inbound cars only. Former London Transport STL1444 is on Eastbourne Street. *Peter Mitchell*

Turning from Everton Road into Fitzclarence Street. *H B Priestley/National Tramway Museum*

Extra from Walton depot on Heyworth Street heading to Old Haymarket for the evening peak. *H B Priestley/National Tramway Museum*

Inbound on the one-way working via Queen's Road and Aubrey Street (seen here). *Peter Mitchell*

192 | THE LEAVING OF LIVERPOOL

Breck Road/Belmont Road. Note the atrocious state of the trackwork.
N N Forbes/National Tramway Museum

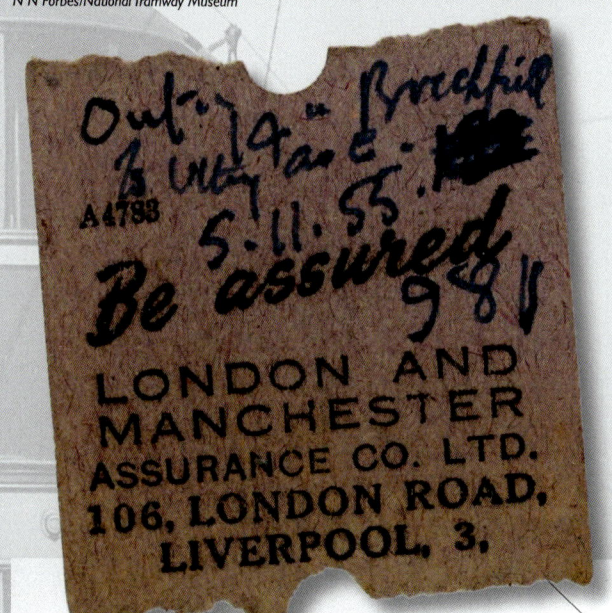

LIVERPOOL CORPORATION PASSENGER TRANSPORT

TRAM ROUTES 13, 14 CONVERSION TO BUS OPERATION

BUS ROUTES 63, 64

REVISION OF ROUTES & NUMBERS

COMMENCING SUNDAY, NOVEMBER 6th, 1955

Journey time	39 minutes (13), 40 minutes (14)
Last day of operation	5 November 1955
Cars seen in service on the last day	158, 181, 182, 201, 205, 212-215, 219, 223, 224, 226, 227, 237, 238, 242-244, 250, 254, 255, 274, 276, 297, 298, 870, 889, 909, 971, 974, 978, 981
Last cars	North John Street south 297 (14), north 219 (14A at 8.02am); 238 (5 Gate, Kirkby, 12.08pm); 182 (13), 158 (14)
Track abandoned	North John Street, Old Haymarket, Islington, Breck Road, Townsend Lane, Townsend Avenue, Utting Avenue East, Lower House Lane. Commutation Row also abandoned.

Breck Road. *Peter Mitchell*

1955 | 193

ROUTE PROFILE
13, 13A, 14, 14A (continued)

Breck Road interlaced track. *H B Priestley/National Tramway Museum*

Townsend Lane narrow single track section. *H B Priestley/National Tramway Museum*

Townsend Lane, end of reserved track inbound. Note outline of former street section, replaced by reservation, in roadway on right. *Peter Mitchell*

Townsend Avenue/Larkhill Lane. *Peter Mitchell*

Townsend Avenue, near Queens Drive. *Peter Mitchell*

Townsend Avenue, Clubmoor railway bridge. *Peter Mitchell*

Utting Avenue East, St Teresa of the Child Jesus church. *Peter Mitchell*

194 | THE LEAVING OF LIVERPOOL

FAREWELL TO THE TRAMS.
They have served you well.

FAREWELL also to:—
RELIABLE SERVICE IN FOG;
CHEAP FARES;
WORKMEN'S TICKETS;
CHILDREN'S HOLIDAY TICKETS;
TRANSFER FACILITIES;
SPEEDY TRANSPORT;
COMFORT;
SAFETY.

Think of these things.
THEY HIT **YOUR** POCKET.
WELCOME TO QUEUES.
The old order changeth.
R.I.P.

Utting Avenue East terminus. In typical Liverpool fashion, 775 is displaying 13A, a route that did not officially serve this location! *Peter Mitchell*

Townsend Avenue. *Peter Mitchell*

East Lancashire Road. *Peter Mitchell*

Lower Lane. *Brian Cook/Online Transport Archive*

A peak hour 13A at Kirkby Admin loop. *H B Priestley/National Tramway Museum*

1955 | 195

Cars withdrawn October-November 1955: 155, 157, 161, 162, 164, 166-170, 175, 177-180, 183-185, 187, 872, 905, 914, 944, 951, 954, 955, 958, 961-963, 966-968, 970, 972, 975, 976, 978 and 979. For a while, 961 served as Edge Lane shunter and 968 was the last with old style fleet numbers. Together with cars withdrawn earlier, all were sold to Maden & McKee who 'reduced' the bodies on the dump so they could be transported by lorry to their scrap yard where they were eventually burnt once saleable material had been removed. This new arrangement followed complaints from residents around Edge Lane objecting to noxious smells coming from the fire dump. Here, 967 is in the process of being 'reduced' whilst 978 and fire-victim 964 are in their fragmented state at Maden & McKee. *[Above and right] Martin Jenkins/ Online Transport Archive*

7 November 1955: Extra morning and evening 6/6A now work to and from Clayton Square. Although Islington was officially abandoned, a disabled car was seen on the inbound track just short of Norton Street. No further information has come to light. The wiring was removed the following day. (JBH)

22 November 1955: For some unknown reason, the wiring on the landward track of the South Loop at Pier Head was removed. This was yet another short-sighted decision by someone at Hatton Garden. As the post-war siding was rarely used, any disabled car had to be pushed onto the truncated link to the former Centre Loop.

13 December 1955: Following an unacceptable build-up of cars in peak hours, the landward track was reinstated.

196 | THE LEAVING OF LIVERPOOL

CHAPTER 13 | 1956

Despite rising fuel costs and the threat to oil shipments from the Middle East, the anti-tram bias within the Corporation continued unabated. The first part of Stage Three of the conversion programme involving routes 19/44 took place on 3 November. Then, on the 22nd, the normally anti-tram *Liverpool Echo* demanded reinstatement of routes 19/44. They even featured the withdrawn cars at Kirkby under the heading "Trams that should be brought back". The paper was inundated with follow-up letters supporting the restoration especially as the buses were consuming an additional 2800 gallons a week. On 27 November, the City Council issued the following statement: "Maintenance levels had been allowed to decline so the trams could not be reinstated." However, no further wiring was removed and the redoubtable local MP Bessie Braddock even raised the matter in the House of Commons but to no avail. Despite this furore, the privately-owned, electrically-powered Overhead Railway was replaced by 42 buses provided by the Corporation.

A few days after the Kirkby conversion, LRTL representative, Ted Gahan, wrote to the Ministry of Transport & Civil Aviation but did not receive this reply until 10 January 1957: "The Regional Transport Commissioner informed the Ministry of Transport that the possibility of reinstating routes 19 and 44 has been considered but has been found impracticable. For example, all the switch gear and overhead cable has been removed and conversion works have been carried out in a tram depot to suit motor bus operation."

December 1955-January 1956: 957 was withdrawn on 13 December 1955 followed by 900, 952, 977 and 986 which survived into early January having been available as Christmas and New Year extras. Some cars had their trucks, buffers and lower panels painted. Also some limited track patching was carried out especially around sunken rail joints.

January 1956: Officially 125 active cars.

Edge Lane	151-154, 156, 160, 165, 201-208, 210-214, 216, 245-249, 261-267, 269, 271-273, 275, 277-280, 283-286, 288, 289, 293, 299, 870, 889, 906, 907, 911, 916, 917, 945, 950, 956, 973, 974, 990, 992
Walton	158, 172, 182, 186, 215, 218-227, 229-232, 235-244, 250-255, 257, 258, 260, 268, 270, 274, 276, 287, 296, 297, 298, 909, 947, 948, 953, 971, 984

3 April 1956: SP3 scrapped.

12 April 1956: SP1 withdrawn. This gaunt Standard stood at the front of the dump for 21 months! Attempts to persuade the Corporation to preserve it ended in failure.

6 May 1956: 168, 170, 177, 181, 183, 188, 946, 952, 958, 962, 963, 968, 972, 977 and SP1 still on the dump. Just two active snowploughs: SP2 at Edge Lane and SP4 at Walton. The former was sometimes used to push withdrawn cars onto the dump.

15 May 1956: A trespasser on a fenced section of reserved track near Pagemoss was struck by 870. This was last fatality involving a tram.

26 May 1956: 246 inside the Works receiving attention to its bodywork; 181, 188, 900 and 978 marked for scrap; and 170, 175, 957, 964, 968 and 972 on the sidings at the back. (JBH)

June 1956: Railgrinder CE&S 234 still making regular outings usually leaving the Works at midnight and returning about 5.30am.

22 June 1956: 175 leaves for Maden & McKee leaving only SP1 on the dump.

15 July 1956: Last LRTL tour using a bogie car. For some reason 186, which was in poor structural condition, was provided. It left the Pier Head at 1.35pm with a full complement of local enthusiasts and visitors and visited Bowring Park and Pagemoss before heading to Kirkby via Scotland Road and Walton Hall Avenue.

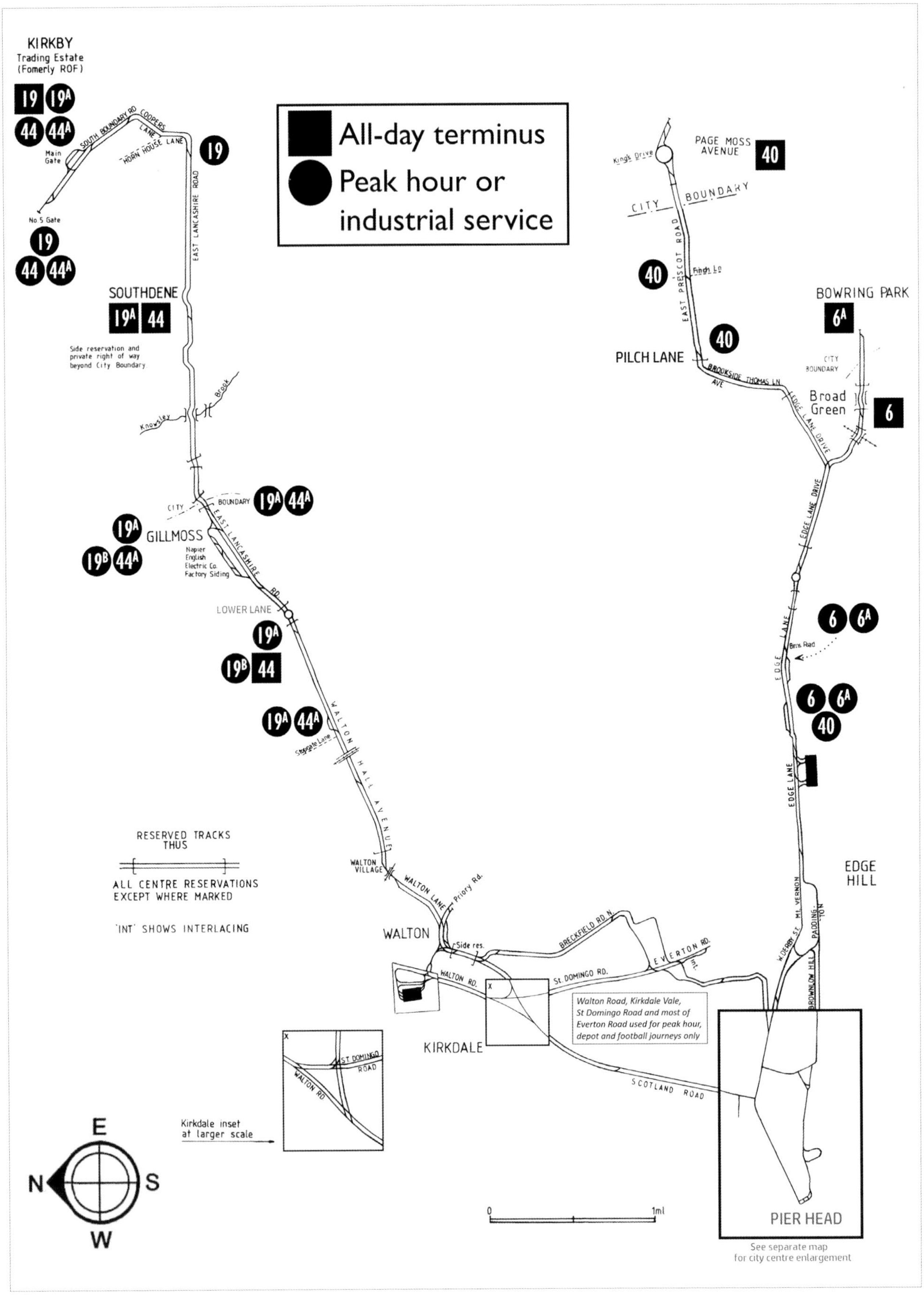

KIRKBY
Trading Estate
(Fomerly ROF)
All-day terminus
Peak hour or
industrial service
PAGE MOSS
AVENUE
SOUTHDENE
BOWRING PARK
PILCH LANE
Broad
Green
GILLMOSS
LOWER LANE
RESERVED TRACKS
THUS
ALL CENTRE RESERVATIONS
EXCEPT WHERE MARKED
'INT' SHOWS INTERLACING
EDGE
HILL
WALTON
KIRKDALE
Walton Road, Kirkdale Vale, St Domingo Road and most of Everton Road used for peak hour, depot and football journeys only
Kirkdale inset
at larger scale
PIER HEAD
See separate map
for city centre enlargement

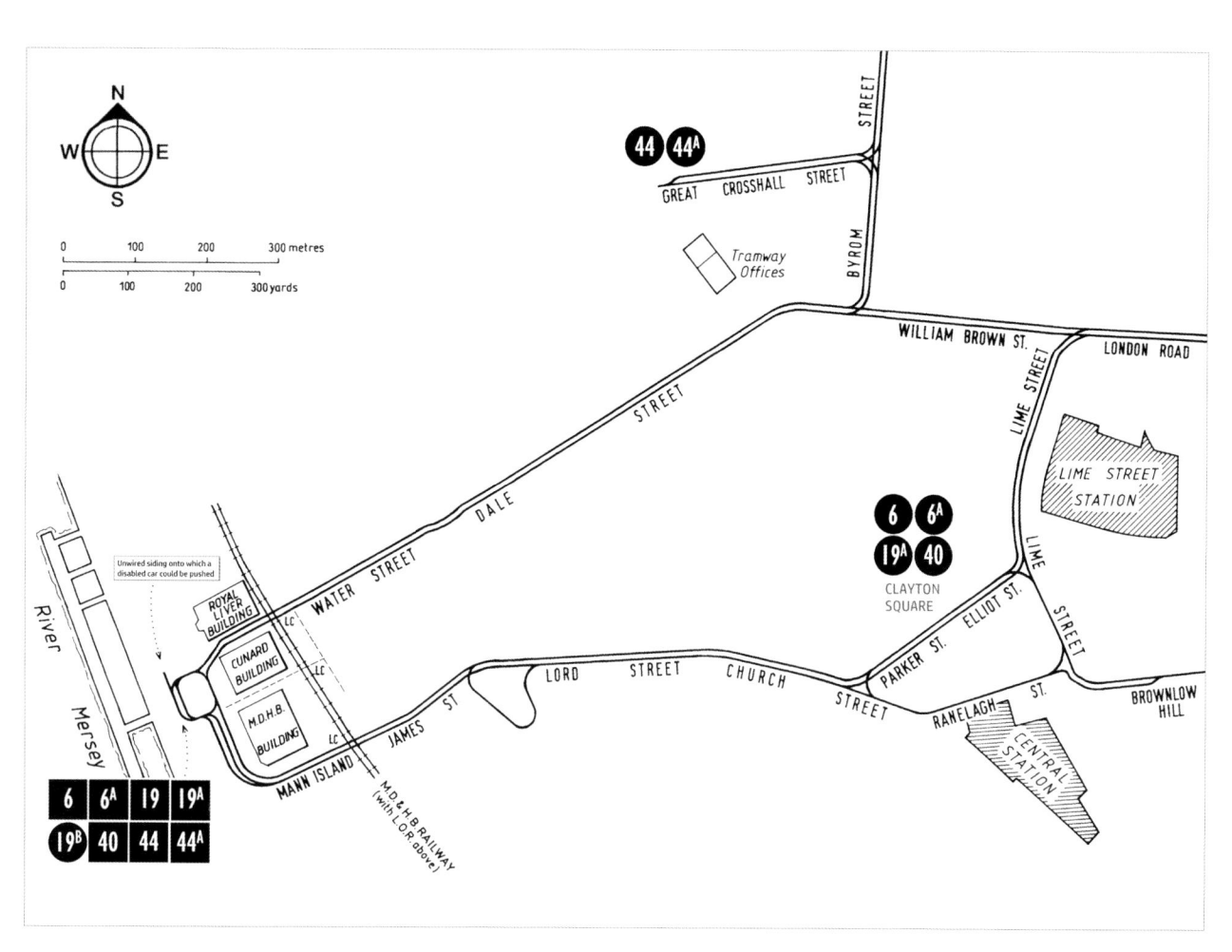

"Sensing this could be his last opportunity to drive a Liner at speed, Harry Tindale has the controller on full parallel wherever possible. Then at the junction with Townsend Lane disaster strikes. A series of ominous thuds and bangs and 186 comes to a swift halt. Outside the overhead vibrates wildly and the car's trolleyhead dangles forlornly at the end of the trolleypole. Not wanting to block the service tracks, people hop out and push 186 down the slight grade to Lower Lane." (MJ) "Everyone proceeded out to push the car but it got the better of us and the result was that it finished up like a marathon with most of us whizzing along on the car and about half a dozen running along after us in our wake." (AFG) *Martin Jenkins/Online Transport Archive*

At Lower Lane, 250 pushed the stricken Liner onto the layover curve on the roundabout, where the trolleyhead was soon refitted by the crew of tower wagon GKD317. Once clear of anyone in authority, Harry was soon back to his best, controller right round, skimming along the rails to Kirkby at high speed.
Martin Jenkins/Online Transport Archive

The tour included a visit to Walton depot, an encounter with an Orange Lodge procession and a visit to Edge Lane scrap yard and depot. In the latter were 285 (smashed front); 947 (burnt out rheostat); 202 (no windscreen) and 244 (no lifeguards, no windscreens and missing panels).

1956 | 199

Mid-August 1956: Transport Department claims Liners consume too much power and cost more to maintain so those at Walton were replaced by Baby Grands from Edge Lane. Although the 31 Liners now confined mostly to peak hour duties they did occasionally appear in all-day traffic.

14 August 1956: Cars on dump: 181, 188, 202, 218, 219, 222, 236, 244, 251, 262, 267, 285, 984, SP1.

17 August 1956: 182 was the last Liner relocated to Edge Lane. The views on the right show Walton-based Liners working on route 19. The first is on the tram-only bridge at Radshaw Nook, the second on Shaw Street. *Peter Mitchell (both)*

15 September 1956: To obtain tape recordings of a handbrake Standard in action, Frank and Ron Oldfield hire SP4 (formerly 646). Leaving Edge Lane at 10.45am it ran as far as Broadgreen. At Edge Lane roundabout, several circuits were made at slow speed in order to maximise the sound effects. Back at the Works, it then ran up and down one of the sidings on the east side. At some unknown date in 1957, it is rumoured it was hired again for filming purposes and is said to have made a return trip to Crown Street but no footage or photographs have ever come to light.
[Below] MTPS collection

3 October 1956: Insufficient serviceable trams. "During the height of the evening peak, Edge Lane depot totally empty, with buses covering for some duties on the 6/6A. Some trams awaiting scrap could easily have been made good." (MJ)

8 October 1956: Cars sold to Bird's of Stratford-upon-Avon were 'reduced' on site before being loaded onto lorries with 181, 188, 202, 212, 218, 219, 236, 244, 251, 262, 267, 276, 285 and 984 having been despatched by 13 October.

200 | THE LEAVING OF LIVERPOOL

"Anxious to retain a tangible reminder of his favourite tram, Brian Martin purchased a dash off 181. Later, I failed to secure the dash from 947, my favourite tram, but I do have one of its interior fleet numbers. Why my favourite? It was the first Liverpool tram I drove. Saturday mid-day at Kirkby sometime in 1955 – no inspectors about – 947 shuttling almost empty between the Admin loop and 5 Gate. I made a couple of return trips. On the last, the crew sat inside chatting. Never forgotten." (MJ)

9 October 1956: 242 and 229 collide in thick fog at the Rotunda.

September-November 1956: To overcome shortages at Walton, an Edge Lane Liner sometimes covered a north end peak hour duty. For example, on 30 October 1956, 916 worked to Pier Head as a 40 before proceeding to Gillmoss (Napier siding) as a 44A. From there it left with a full load working as far as Lime Street as a 19A after which it returned to Edge Lane by way of Brownlow Hill. The car was crewed throughout by Edge Lane staff. It is not known which Liner undertook the last of these fill-in duties.

> Townsend crossing, Lower Lane,
> Rain against the window pane,
> Ceiling leaking, woodwork creaking,
> On a tram to Kirkby.
>
> Round the island, off again,
> Race a bus to Stonebridge Lane,
> Windows streaming, driver dreaming,
> On a tram to Kirkby.
>
> Gillmoss siding, Croxteth Brook,
> Cross the bridge to Radshaw Nook,
> Eight wheels dragging, body sagging,
> On a tram to Kirkby,
>
> Southdene cutting, then the straight,
> To the curve where 'in' cars wait,
> Motors wailing, lifeguard trailing,
> On a tram to Kirkby.
>
> Hornhouse Lane, a left hand swerve,
> (Signal lights announce the curve),
> Acceleration, what elation,
> On a tram to Kirkby.
>
> Past the Admin, round the loop,
> See how much the platforms droop,
> Compressor humming, well worth coming,
> On a tram to Kirkby.

This poem, penned by enthusiast Don Littler, perfectly captures a ride to Kirkby on a Liner on a wet day.

27 October 1956: Showing just 'Breck Road' on its screens, 204 operated the city's last ever football special from Carisbrooke Road to Grant Gardens at about 5pm following an Everton FC home game. Until quite recently, football specials had left Walton depot and run via Kirkdale Vale and Walton Lane in order to reverse on the football spur in Priory Road, after which they lined up bumper to bumper on Walton Lane siding awaiting the exodus from Goodison Park. 261 and 277 are seen on the Priory Road stub on 13 October 1956. *[Above] Martin Jenkins/Online Transport Archive*

These specials ferried fans to the Pier Head via either Lime Street or Scotland Road. 271 is seen edging forward to the pick up point on Walton Lane in 1954. *[Below] J G Parkinson/Online Transport Archive*

28 October 1956: The last Liner round the South Castle Street loop was 870. "Sparks usually fly each Sunday as the wheels bite into the little used rail. No one knows how this solitary once-a-week extra catering for visitors to Broadgreen Hospital has survived. Latterly, former points were locked in position and rails are filled with dust and grit." (MJ) This view of 990, working the hospital special, dates from earlier in the month. *[Below] Martin Jenkins/Online Transport Archive*

TRAM 19, 19ᴬ, 44 ROUTES

CONVERSION TO BUSES

Commencing SUNDAY, 4th NOVEMBER, 1956

THE ABOVE SERVICES WILL BE CONVERTED TO BUS OPERATION WITH NEW TIME-TABLES AND ROUTE ALTERATIONS AS FOLLOWS:-

ROUTE 19 – AS AT PRESENT
ROUTE 19ᴬ – WILL BE EXTENDED via MOORGATE ROAD TO KIRKBY INDUSTRIAL ESTATE

ROUTE 44 – WILL BE EXTENDED via MOORGATE ROAD, BEWLEY DRIVE, BROAD LANE, PARK BROW DRIVE, WESTHEAD AVENUE, MINSTEAD AVENUE to OLD ROUGH LANE (Northwood) AND RE-NUMBERED 44ᴬ

FOR DETAILS SEE HANDBILLS AVAILABLE AT ENQUIRY OFFICES

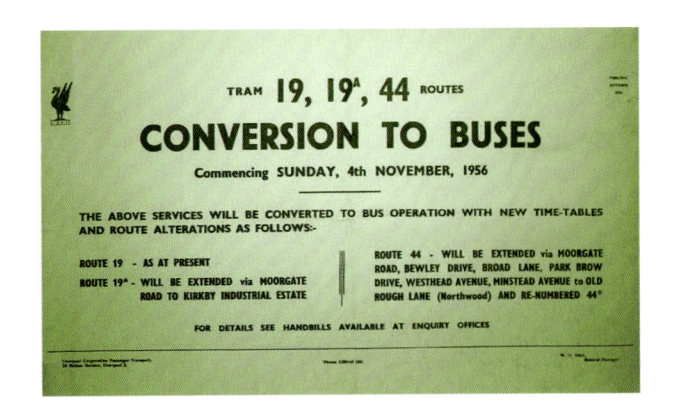

3 November 1956: A black day with over 15½ route miles abandoned including much track opened in the late 1930s/early 1940s.

The 9¾ mile route 19 linking the Pier Head to Kirkby via Church Street, Lime Street, London Road, Shaw Street, Robson Street, Walton Hall Avenue, Lower Lane, East Lancashire Road, South Boundary Road and the short working 19A to Southdene were replaced. The numbers 19/19A/19B were used for a range of short workings often because conductors failed to display the correct number or, at other times, cars showed only a route number or destination. Inbound short-workings included Lower Lane, Everton Valley (for Walton depot), Breck Road (for Grant Gardens and Walton depot), Lime Street (sometimes used for cars working to Clayton Square) and Clayton Square. Outbound cars turned short at Breckfield Road North, Robson Street, Walton Lane, Priory Road (often blank on blinds), Stopgate Lane (cars mostly showed just 'Stopgate Lane' with no route number), East Lancashire Road (sometimes Lower Lane, rarely Lower House Lane), Gillmoss (shown by cars using the crossover at Croxteth Brook as well those accessing the Napier siding), Southdene, Hornhouse Lane, Kirkby and 5 Gate (shown on blinds as Kirkby). In peak hours, some cars leaving Walton depot ran first to Grant Gardens where they reversed in order to proceed north via the 19. Throughout the week, the 19 ran every 20 minutes whilst the 19A was every 10 minutes except on Sunday mornings when it dropped to 20 minutes.

The 9½ mile 44 linking Pier Head to Southdene via Scotland Road, Everton Valley, Walton Hall Avenue and Lower Lane was replaced together with the 44A, which was used for a variety of short workings. Working towards the city, these included Spellow Lane (for Walton depot), Rotunda and Great Crosshall Street. In another of those Liverpool quirks, inbound 44/44As running through to Pier Head usually showed 'via Dale Street' whilst those terminating at Great Crosshall Street 'via Scotland Road'.

Cars leaving the city usually showed 'Spellow Lane' when returning to depot. Other short workings included Walton Lane, Priory Road (usually blank on screens), Stopgate Lane, Lower Lane (sometimes East Lancashire Rd or Norris Green/Lower Lane on screens) and Gillmoss (for Napier siding and Croxteth Brook). Outbound 44/44As from Pier Head usually showed 'via Dale Street' but outbound from Great Crosshall Street 'via Scotland Road'. Quite a few journeys on the 44/44A were extended to both Kirkby Admin Gate and 5 Gate in the morning and evening peaks as well as on Saturdays. Service frequencies on the 44 varied between every 10 and 15 minutes except for most of Sunday when it was every 20 minutes.

Last Liners seen in service (6/6A/40): 151, 158, 160, 182, 909, 947, 950, 956, 974. Amazingly, 'mobile ruin' 950 still active. 151 made the final passenger trip on a mid-day peak extra from Broadgreen to Crown Street. Here the screen was changed to 'Edge Lane' and the tram proceeded to the shed, off-loading its passengers and disappearing inside. "I had ridden the penultimate Liner 182 and was waiting outside Edge Lane for 151 to return. When it did, I nipped on board and bought a' 1d ticket from the conductor – the last ever purchased on a Liner." (MJ)

Last cars: The number of the last 6A to leave the Pier Head for Bowring Park via Dale Street is not known.

"The last 19 was relatively empty when it left the Pier Head at 11.05pm. Although carrying the chalk inscription 'Last 19' on its side, 206 attracted little attention as it gradually filled with passengers along Lord Street and at Clayton Square. It seemed just like any other night with the brightly lit car stopping to let people on and off. However, by the time we reached Lower Lane only a handful of non-enthusiasts remained. After snaking and lurching along the dark stretches of the East Lancashire Road and South Boundary Road, we reached Kirkby at around midnight. All 20 or so enthusiasts alighted and a few of us lined up to take precious flashlight photographs. We had to stand in line in the hope at least one flash bulb would actually go off – they were very temperamental! Driver Hughes was clearly 'upset' and told us he was transferring to Edge Lane so he could keep working on the trams – 'too old for buses'.

"After we piled on board most of us up front behind the driver as 206 fairly galloped along until we reached Southdene. Here most of us alighted, waiting for the last 19A. We watched the red tail light on 206 recede into the distance. It was chilly and pitch black. We were huddled at the isolated crossover installed in 1953. Then piercing the gloom we heard the distant whine of motors and the squeal of wheels as 207 lurched round the curve at Moorgate Road. Here was the last 19A still with the inscriptions I had chalked on its sides early in the evening. It was now about 12.50am. After using the crossover and changing the blinds to show 'Spellow Lane', 207 set off carrying passengers for the last time over the magnificent reserved tracks along East Lancashire Road and Walton Hall Avenue. After a typical spirited late night run, we reached the depot about 1.10am. We stood and watched as 207 was guided inside by the shunter. It was hard to accept it was all over." **(MJ)** [Opposite, top]

Earlier that night, 293 operated the final 44 which left the Pier Head at 11.40pm and arrived at Southdene just before 12.30am. A few enthusiasts opted to make ride this rather than the last 19 and 19A.

Walton depot: (See page 117 for depot plan.) In its heyday, 'Spellow' had about 120 cars, some of which were stored in the open at night. By the end just the five centre roads in the

1901 building remained. Originally, there were three groups of five roads each accessed from Harlech Street by way of an arched entrance. The former horse car depot accessed from Carisbrooke Road had stopped housing service trams in late 1951 but was then used for a while as a store for works cars and withdrawn trams. It is believed the Carisbrooke Road siding continued in use until late 1951. 270 is seen entering the surviving arch sometime in 1956. Note the Tramway pub on the right, now a convenience store. The depot finally closed in 1989.

[Right] Martin Jenkins collection/Online Transport Archive

4 November 1956: Remaining trams now confined to Church Street as Dale Street was closed in preparation for a one-way traffic scheme.

ROUTE PROFILE
19, 19A, 19B, 44, 44A

The first images cover the 19s as far as the junction with the 44 at Anfield Road/Walton Lane.

Church Street. Phil Tatt/Online Transport Archive

Turning from Lime Street into London Road. R W A Jones/Online Transport Archive

Everton Road. H B Priestley/National Tramway Museum

Breckfield Road North. Peter Mitchell

204 | THE LEAVING OF LIVERPOOL

These images follow the 44 to the junction with the 19 at Anfield Road/Walton Lane.

New Water Street with the Liverpool Overhead Railway structure in the background and an unusual triangular-shaped tram stop. *E A Gahan/Online Transport Archive*

Tram Route 44								KIRKBY—CITY	
Stage No.									Stage No.
6								Kirkby Estate	25
8	1½							Ormskirk Road	23
10	2½	1½						Ainsworth Lane	21
12	4	2½	1½					Radshaw Nook	19
14	4	2½	1½	1½				Gillmoss (Stonebridge L./Back Gillmoss L.)	17
16	5	4	2½	1½	1½			Lowerhouse Lane	13
20	5	5	4	2½	1½	1½		Townsend Avenue	11
22	6	5	4	4	2½	1½	1½	Stanley Park Avenue	9
24	6	6	5	5	4	2½	1½	Bodmin Road	7
26	6	6	5	5	4	2½	1½	Walton Breck Road	5
28	7	6	6	6	5	4	2½	Rotunda	3
30	7	7	6	6	6	5	4	Richmond Row	1
	7	7	7	6	6	5	4	2½ 1½ City	

Byrom Street. *Peter Mitchell*

1956 | 205

ROUTE PROFILE
19, 19A, 19B, 44, 44A (continued)

Great Crosshall Street, peak hour only terminus. *Peter Mitchell*

Scotland Road. *Peter Mitchell*

Bottom of Everton Valley. *H B Priestley/National Tramway Museum*

Kirkdale Vale, peak hour car returning to Spellow Lane on depot-only track. *Martin Jenkins/Online Transport Archive*

Walton Road, car returning to Spellow Lane along depot-only track. *Peter Mitchell*

Top of Everton Valley. *Peter Mitchell*

TRAM 19, 19ᴬ, 44 ROUTES
CONVERSION TO BUSES
Commencing SUNDAY, 4th NOVEMBER, 1956

THE ABOVE SERVICES WILL BE OPERATED BY BUSES INSTEAD OF TRAMS WITH NEW TIME-TABLES AND ROUTE ALTERATIONS AS FOLLOWS:—

ROUTE 19 - AS AT PRESENT

ROUTE 19ᴬ - WILL BE EXTENDED via MOORGATE ROAD TO KIRKBY INDUSTRIAL ESTATE

ROUTE 44 - WILL BE EXTENDED via MOORGATE ROAD, BEWLEY DRIVE, BROAD LANE, PARK BROW DRIVE, WESTHEAD AVENUE, MINSTEAD AVENUE to OLD ROUGH LANE (Northwood) AND RE-NUMBERED 44ᴰ

FOR DETAILS SEE HANDBILLS AVAILABLE AT ENQUIRY OFFICES

Liverpool Corporation Passenger Transport, 24 Hatton Garden, Liverpool 3.

'Phone CENtral 7411

W. M. HALL, General Manager

This final sequence shows the combined route from Walton to Kirkby.

Anfield Road/Walton Lane. *R B Parr/National Tramway Museum*

Walton Lane/Spellow Lane. *E A Gahan/Online Transport Archive*

Walton Lane with football siding on the right for Everton FC. *N N Forbes/National Tramway Museum*

Walton Lane, approaching the narrow single track section under the Bootle Branch. *H B Priestley/National Tramway Museum*

The twisting track through Walton Village. *H B Priestley/National Tramway Museum*

Walton Hall Avenue/Queens Drive. *J S Webb*

Walton Hall Avenue with the CLC railway bridge in the background. *Peter Mitchell*

1956 | 207

ROUTE PROFILE
19, 19A, 19B, 44, 44A *(continued)*

Stopgate Lane. This was one of only two roundabouts on the system where trams passed through the middle. *N N Forbes/National Tramway Museum*

Stopgate Lane, including peak hour siding, serving the Long Lane industrial complex. *N N Forbes/National Tramway Museum*

Walton Hall Avenue. *Peter Mitchell*

The wartime, long double-track Napier siding. *Pam Eaton*

East Lancashire Road/Gillmoss Estate. *R J S Wiseman/National Tramway Museum*

208 | THE LEAVING OF LIVERPOOL

East Lancashire Road, on the fast section without stops between Southdene and Hornhouse Lane, passing the ICI Metals factory. *Alan A Jackson*

Hornhouse Lane. *J H Roberts/Online Transport Archive*

South Boundary Road, Kirkby, with Cooper's Farm in the background. *J S Webb*

5 Gate. *Peter Mitchell*

Journey time	55 minutes (19), 49 minutes (19A), 48 minutes (44/44A)
Last day of operation	3 November 1956
Cars seen in service on the last day	205-208, 210, 213, 215, 221, 224, 226, 231, 232, 237, 238, 240, 247, 250, 252-255, 257, 258 260, 268, 270, 271, 274, 275, 283, 287, 293, 296, 297-299
Last cars	Last peak hour cars: 214 (5 Gate), 254 (Gillmoss sidings), 240 (Great Crosshall Street), 240 (44A), 226 (44 from Kirkby), 204 (44 from Grant Gardens). Last service cars: 206 (19), 207 (19A), 293 (44).
Track abandoned	Water Street, Dale Street, William Brown Street, Byrom Street, Scotland Road, Kirkdale Road, Walton Road, Harlech Street, Carisbrooke Road, Spellow Lane (disused), Kirkdale Vale, St Domingo Road, Everton Road, parts of London Road, Moss Street, Shaw Street, Eastbourne Street, Village Street, Fitzclarence Street, Queens Road, Aubrey Street, Breck Road, Breckfield Road North, Robson Street, Sleepers Hill, Everton Valley, Walton Lane, Priory Road siding, Walton Hall Avenue, East Lancashire Road, Hornhouse Lane, Coopers Lane, South Boundary Road

During the morning of 4 November, Walton's 42 Baby Grands were transferred to Edge Lane in several small convoys travelling via Walton Road, Kirkdale Vale and Heyworth Street where they used tracks of former route 19 as far as Lime Street, after which they followed route 40. "Last ten: 240, 214, 232, 231, xxx, xxx, 283, 271 (pause) 287 (pause five minutes) then 221, last of all showing 10B Pier Head on rear blinds." (JBH) With arrival of Walton's Baby Grands, Edge Lane depot released 261, 277 and 286 for scrap which joined 211, 216, 220, 242 and 278 on the bank.

Also during the early part of the day, the remaining 31 Liners were driven from Edge Lane to Kirkby, again in convoys, probably making the cross-city journey by way of route 40 to Church Street and then by former route 19 from Clayton Square. It is not known whether these transfers were undertaken by drivers from Walton and/or Edge Lane or whether shed and works staff were also involved. It was certainly the last time many would drive a tram. 160 and 974 are nearing the end of their journey. Everything was completed by 1pm with 153 making the final one-way trip from Edge Lane. *[Below] Stan Watkins*

The abandonment of 3 November 1956 led to the closure of the tracks in William Brown Street, Dale Street, Water Street and New Water Street. The first view shows an inbound 6 descending William Brown Street towards the junction with Byrom Street and the second, taken from the Overhead Railway, is looking up Water Street towards the Town Hall. *[Above] H B Priestley/ National Tramway Museum; R J S Wiseman/National Tramway Museum*

Crew time boards were produced by hand in the Schedules Office on the second floor at 24 Hatton Garden. This is the only one known to have survived and covers a split duty on the 6/6A with C standing for Church Street and D for Dale Street. It is interesting to note that the 24 hour clock is used, although this was never used in public documents in the tramway era.

[Below] National Museums Liverpool

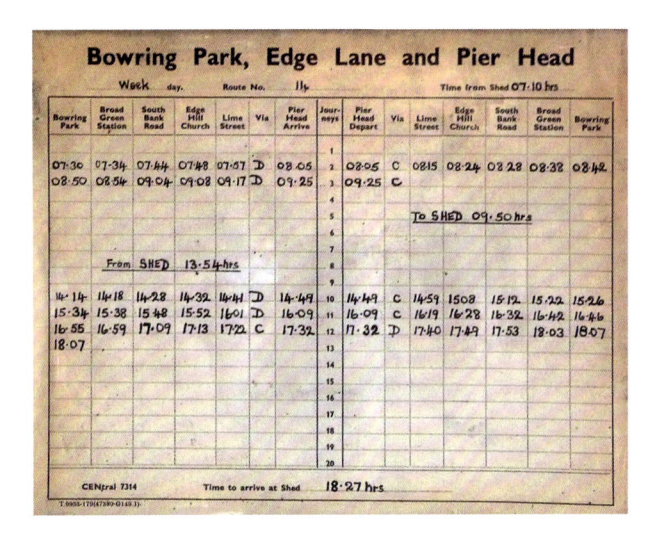

On arrival, the cars were parked on both tracks of the 5 Gate extension. This view taken towards dusk on the Sunday evening shows 186, 971, 950, 154, 165, 182, 172, 909, 907, 870, 973, 156, 158, 947 and 956. By nightfall trolleys were tied down and doors closed against intruders. *[Opposite, top] Anthony Henry*

However, 153 remained active as a shunter and mobile 'headquarters' for the night watchmen. It had been rebuilt in 1950 and repainted in July 1955.

"A small group of us befriended the watchmen on 153 and with the help of a couple of packets of cigarettes persuaded them to take us for a spin almost as far as Hornhouse Lane. From memory, we had no lights on so as not to attract attention. After swinging the trolley in total darkness, 153 completes its illicit run on the same track. This is probably the last ever move made over any distance by a Liner in Liverpool. For a few more days, cars made their way to the railway crossing under their own power until the current was switched off on 9 November 1956. It is rumoured that Manager Hall also came to Kirkby for a turn at the controls." **(MJ)**

[Opposite, centre left] Martin Jenkins/Online Transport Archive

210 | THE LEAVING OF LIVERPOOL

5 November 1956: Work began on moving the redundant trams within Kirkby Industrial Estate, onto railway sidings which had once served a wartime meat packing factory. Here, the trams could be burnt without 'offending' those living near Edge Lane. Each tram was driven under its own power to the crossing with the internal rail system where, with the aid of steel plates to reduce the lateral friction, it was tugged and pulled through 90 degrees onto the railway. The heavy work was done by a Leyland Beaver 'Jack' Wagon chained to the front with a Corporation breakdown tender at the rear to ensure the tram did not topple over. The Leyland was subsequently preserved. *[Above, right] G C Bird/Online Transport Archive*

Once on the internal railway system, the cars were towed for about ¼ mile in convoys of four by one of the Corporation's 0-6-0 Vulcan-Drewry diesel shunters. "Cars are loose coupled with chains, a man being on each car to work the handbrake." (JBH) *[Overleaf, top] Leo Quinn collection/Online Transport Archive*

During these moves, 992 suffered a broken truck and had to be repaired! On reaching the sidings, the stranded Liners were locked and bolted except for 186 which was used by the night watchman.

10 November 1956: The final movement occurred when diesel shunter No 4 pushed 186 in front of 907.

1956 | 211

10 December 1956: In response to the worsening fuel crisis, the Corporation reluctantly introduced a 10 minute headway on the 40 which required eight instead of four cars off peak. However, in a totally illogical move aimed at saving fuel, late night trams were curtailed along with the buses! After the 9.55pm departure from the Pier Head, all 40s terminated at Pilch Lane.

However, there was some track patching plus very limited relaying using rail lifted from other locations. The rail grinders were still at work in the early hours.

11 December 1956: To have sufficient serviceable trams, 223 went into the Works to be overhauled and fitted with new tyres.

25 December 1956: Merseyside was enveloped by thick fog and snow. SP4 is believed to have undertaken snow clearing duties at 3am. CE&S 273 also ventured out carrying bags of salt to keep the points clear.

27 December 1956: SP4 out early morning ploughing reserved track sections.

30 December 1956: Despite the fuel crisis, the unique Liverpool Overhead Railway was closed with the Corporation providing the replacing buses. *[Right] Phil Tatt/Online Transport Archive*

LIVERPOOL CORPORATION PASSENGER TRANSPORT

ROUTE
PAGE MOSS **40** PIER HEAD

INCREASED SERVICE
(During Fuel Shortage)
COMMENCING
MONDAY, 10th DECEMBER, 1956

212 | THE LEAVING OF LIVERPOOL

CHAPTER 14 | 1957

At the beginning of the year, 66 Baby Grands at Edge Lane, 52 of which were required to cover duties on the 6, 6A and 40. Also serviceable were SP2 and SP4 and work cars CE&S 234, 273, 287.

8 January 1957: Visit to the Works. 226 (new tyres), 214 (off its trucks), 204/216 (in body shop). Also on site SP2 and CE&S 234 and 287. In an attempt to deter souvenir hunters, it was increasingly difficult to access the scrap areas.

January 1957: The South Loop inside track again dewired. Following cars had their lower panels repainted during the early months of the year: 204, 207, 210, 213, 215, 231, 237, 249, 250, 252, 253, 257, 260, 266, 272, 279, 298.

17 January 1957: SP4 had a brake test – just in case!

22 January 1957: Owing to the ongoing fuel crisis, the local *Evening Express* stated trams would continue to January 1958. 204, 227 and 248 returned to service.

February 1957: Some minor track work at Edge Lane roundabout. Scrapping by George Cohen & Sons started at Kirkby.

8 February 1957: All too often trams were taken out of service due to minor faults, bumps or scrapes. However, 245 and 265 were retyred, rewired and given fairly thorough overhauls and then a miracle! After having been taken off the dump in early January, collision victim 216 returned to service with newly painted lower panels and is seen on 8 September 1957 about to leave the single track at the foot of Brownlow Hill on a section of line opened just 20 years previously. *[Below] A D Packer*

March 1957: A shortage of buses, coupled with a surplus of tram drivers nearing retirement age, offered the prospect of the final conversion being extended into 1958 but Hatton Garden remained committed to completing the programme claiming some track was dangerous. 206, 221 and 239 withdrawn but 270 returned to service after being laid up for four months.

2 April 1957: Transport Department stated trams would remain until at least September. A local press headline read "Trams Reprieved – we would not be justified taking them off whilst petrol rationing continues". However, no more trams had their lower panels repainted whilst 284 became the last to be fitted with new tyres.

25 April 1957: The last liner to be burnt at Kirkby, 186 had been reconditioned in 1951 and 1955.

"It looked very forlorn surrounded by piled wood like a Viking funeral pyre. More and more enthusiasts visited the Kirkby graveyard at weekends to secure items such as seats, side indicators, fleet numbers, pressure gauges, bells, strap hangers, rear lights and lampshades some of which survive today." **(MJ)** *[Below] Martin Jenkins/Online Transport Archive*

29 April 1957: As the Suez crisis eased, route 40 reverted to every 20 minutes except on Saturdays and at peak times. Late night bus and tram services were also restored. By now maintenance was virtually non-existent. Some surviving 'runners' such as 210, 263, 272, 279, 280 were in really bad condition. 263 looks very down at heel as it makes its way along the reserved track on East Prescot Road on 25 June 1957. *[Right, top] Peter Mitchell*

20 May 1957: A new one-way traffic scheme around St George's Hall and the Mersey Tunnel came into force and, in a surprise move, a loading island was built on Lime Street for outbound cars.

June 1957: Newly recruited 'clippie', Sheila, was assigned to the trams.

2 July 1957: On the dump: SP1, SP4, 206, 211, 220, 221, 229, 242, 243, 247, 261, 265, 277, 278, CE&S 283 and 286. SP2 still acting as yard shunter. 261 still had the old-style gold fleet numerals both upstairs and downstairs.

"One shed foreman was aggressively unfriendly but there was one who would still let us in. Basically he turned a blind eye. During these visits, items such as fleet numbers and indicator blinds were removed and smuggled out. Fortunately, one of our number was 6ft 4in tall and was able to walk out (rather oddly!) with a blind concealed down each trouser leg." **(MJ)**

26 July 1957: 239 dumped following a minor accident.

12 July 1957: Owing to Orange Lodge Processions, 6s diverted via Brownlow Hill.

7 July 1957: 271 made the final circuit of the Victoria Monument on the Sunday only working to Broadgreen Hospital. When the special turned up the following week, the driver encountered a length of new pavement blocking the outer rail and had to reverse onto Lord Street in order to run down to the Pier Head.

In his detailed notebooks, Tony Gahan recorded the names of many drivers and conductors based at Edge Lane: Harry Tindale, Bobby Mitchell, Walter Benstead, G Prosser, Fritz Pullman, Jack Somers, Tommy Kirkham, Albert (Olly) Wilson, Tom Webster (senior driver, 38 years' service, and last driving instructor), Jim Minnen, Alf Kitchin, ? Noonan, H Evans, William Evans, Jim Minnery, Jim McKeeley, Stan Morris, Eddie Baily, John Naughton, Bill Prosser, Jim Scott, Bob Norman, G Rodden, George Taylor. Tony also listed his own nicknames for some others – the origins for some are quite obvious with others lost in the mists of time: Teddy Boy, Tall Boy, Athlete, Rastus, Poncho, Tucker 1, Hangman, Ears, King George VI, Fuzzy, Slim Whitman, Mad Lad, Fast Fellow, False Teeth, Tucker 2, Blue Nell, Smiley, Tummy, Pete, Paddy, Skull, Gummy, Sailor Sam, Binman, Chinee, Taffy. Among the conductresses were Nellie Wallace, Rose Robinson, Tessie and Sheila,

Early August 1957: All withdrawn cars sold to Cohens. SP4 broken up by 7 August. Last car, other than 293, to receive attention at the Works was 272.

8 August 1957: Overhead wires brought down when a Crosville bus skidded and collided with a traction pole outside Edge Lane Police Station. Then a second Crosville became entangled in the fallen wires causing more damage. The tram service was curtailed at Durning Road crossover with connecting buses into town. This was the first time there had been no trams in the city centre since the worst days of 1941 blitz. By early afternoon, the overhead had been repaired and the tram service restored.

22 August 1957: "Front controller on 214 fails. With driver using the rear controller, car is driven with passengers all the way from Lime Street to depot with the conductor monitoring progress from the front platform. 270 rides like a bus. Trolley wheel drums very noisily passing under ears and frogs." (AFG)

23 August 1957: Final track 'repairs' – protruding setts levelled but dished rail joints left untouched.

2 September 1957: Some drivers having a final fling. 272 completed journey from Pagemoss to the depot in just 9 minutes – "absolutely 100%". (AFG)

Early September 1957: Following cars on the bank: 230, 232 (collision), 240, 243, 247, 255, 268 and 283. The latter had only been rebuilt in 1955. Some were more structurally sound than surviving 'runners'. "In fact, a few were virtually limping out of the depot, bodies bent and bowed." (AFG)

> "I recall a fire breaking out in the cupboard under the stairs on 205 as it made its way along Church Street. Everyone turfed off and 205 seen later at the Pier Head on the siding quietly smouldering. Drivers require a lot of skill to simply get through a day's shift. On wet days, water leaking in round the cab windows drips onto the controllers. Most have wodges of paper jammed into the gaps in the window surrounds. Majority of drivers wear gloves. I remember when the hand-operated windscreen wiper on 263 failed to turn, the driver was forced to stand his face pressed to the window to see where he was going." **(MJ)**

8 September 1957: At 6.20am on the Sunday morning 'Liverpool's Last Tram' entered service on the 6A. Historically, 293 had been rebuilt in early 1952. Then it went into the Works in January 1956 when it was repainted and given the truck from 247. It re-entered service on 24 February 1956. However, it was back in the Works just over a year later following a collision. At some point, it was selected to be the last tram. It was fitted with new ventilators and beading to stop leaks as well as refurbished seats and smaller, silver painted lifeguards. Its motors were also overhauled. Following trial runs around the Works perimeter on 20 August, it was officially returned to the active fleet on Friday 6 September. It looked very smart in its reverse cream and green livery complete with last tram inscriptions and proved popular with the public, some even thinking it was a 'new tram'. It was out every day during the final week mostly on the 6A but on Tuesday 10 and Thursday 12 September it was assigned to the 40.
[Left, bottom] J B C McCann/Online Transport Archive

Harry Tindale (who else?) was again in charge of the final LRTL tour on board car 245. The well-filled car left the Pier Head and traversed all remaining track including the virtually disused Liverpool Road siding at Pagemoss. It also completed three circuits of Edge Lane roundabout. On two occasions, it posed alongside 293. [Above] J H Price/National Tramway Museum

The car carried home-made posters on the side and among the enthusiasts seen in the lower saloon are Anthony Henry, John Horne and John Price. [Above] Leo Quinn/Online Transport Archive

At the end of the tour, there was an emotional 'farewell' as 245 finally entered the depot. "The last Liverpool tour, like the first eighteen years before, was a fast ride on a fast modern car over a fast system – that is how it will always be remembered." (Modern Tramway)

10 September 1957: 293 was spotted fully-laden on the 40 but with two conductors. This was probably due to staff taking time to issue the range of old-style, Bell Punch 'Last Tram Week' tickets. Secure in their tightly-wired ticket racks, these 'souvenirs' were carried by all conductors during the final week. The set included a full range of workmen's returns as well as the long-established child 1d return, both of which concessions were coming to an end. Unfortunately, the tickets did not make provision for transfers between the 6A and 40 which caused some confusion.

It is possible that one of the two conductors was Bill Peters who had made a special request to conduct on the trams and had been transferred from PAR for the day. He has written an account of his experiences in his autobiography, *Busman*. Note that the term 'guard' was frequently used in Liverpool rather than 'conductor'.

> "I had received brief instructions for guarding on a tram which was rather more involved than a bus. To do the job properly, when he's got his fares in, the guard remains on the platform holding the trolley rope so that if the pole dewires it can be held down to prevent damage to the overhead lines. The tram looked enormous, the top deck of forty seats seemed endless after a 30 seat upper deck on a bus. I wondered if I would be able to get round the fares, especially with the strange gear [rack of tickets and an ex-London punch]. The ceiling was stained with nicotine, the leather upholstery rotting, and the chrome pitted with rust. It looked rather sad.
>
> "It was another world. Everything was more relaxed [than the buses], more leisurely, even the passengers were different. We went round the loop at the Pier Head but at Bowring Park seat backs had to be reversed and there was a double set of indicators to be changed. The trolley pole had to be turned too and at the first attempt I tried to walk alongside the car in a straight line, but the tension in the rope and a powerful upward pull soon reminded me to walk in a wide semi-circle. We ran in on finishing, unusually for an early, and the last job was changing the facing points to the shed. I said good-bye to my driver with the usual back-handed compliments and went to pay in. We had carried 620 passengers and taken eleven pounds odd – a modest day's work for 1957."
> **(Bill Peters)**

Cars for the final procession to be selected from 207, 210, 213, 214, 223, 226, 235, 245, 252, 257, 260, 264, 266, 269, 296 all of which had had their bumpers and couplers painted. Overlooked by Liverpool University's Victoria Building on Brownlow Hill, 257 approaches the junction with Crown Street on 25 June 1957. The buildings on the right have since been demolished.
[Right, top] Peter Mitchell

201, 203, 204, 205, 208, 241, 248, 249, 250, 253, 270, 273, 275, 279, 284, 287, 289, 297 and 299 were withdrawn before the last day some having operated during the Friday evening peak on 13 September. Dating from 1942, 299 was the newest in the fleet and is seen on Edge Lane on 25 June 1957. It was unusual in having numbers such as 6B and 6C on its screens. Note the colourful array of advertisements on the left. *[Above] Peter Mitchell*

13 September 1957: "Straight over after school. Squeeze in as much riding as possible – 269, 224, 288, 271, 238. Must record the nocturnal voyages of the last evening cars in NORMAL passenger service. Flashlight of 238 at PH. Fantastic moment at Bowring Park. After talking to the crew and taking a flashlight photo, the driver asks if I would like to take 245 over the crossover. Bit nervous. Couple of notches and over safely. How about up the hill? Yes please. To Broadgreen? Yes. Stop – another flashlight with conductor, George Taylor. On to the junction with the 40 – then the roundabout – seventh heaven. Finally just short of the depot I handover. The driver smiles – sadly. 'I shall be sorry to see them go – better than these replacing buses.' I agree. Upstairs front seat for last night time trip to PH. Goodbye to the crew. Ferry boat and back home." **(MJ)**

[Opposite] Martin Jenkins/Online Transport Archive

Later that night, 245 worked the midnight departure on the 6A, believed to have been driven by Alf Kitchin.

14 September 1957: Stage Three of the conversion programme ended with replacement of the 6, 6A and 40. The 6s departed from the remnants of the South Loop and, during the

216 | THE LEAVING OF LIVERPOOL

course of their five mile run, passed through the city by way of Lord Street, Church Street, Clayton Square, Lime Street, London Road and Pembroke Place. Here they encountered a single-track, one way section with outbound cars using Mount Vernon and inbound cars Paddington and Crown Street. The main section was along Edge Lane and Bowring Park Road with 6s terminating at Broadgreen (sometimes Broad Green Stn or Broadgreen Stn) and 6As at Bowring Park, the termini being referred to by crews as 'The Green' and 'The Park'. Outside peak periods, there was a 10 minute service at most times.

The five-mile 40 was a relatively new route dating from June 1937. In the central area it diverged from the 6s to use Brownlow Hill before accessing the same one-way system used by the 6s. The two routes then followed the same path until the 40s branched off at Gardeners Arms along Thomas Lane and Brookside Avenue. From Knotty Ash, they reached Pagemoss by way of the grass tracks along East Prescot Road. There was a 20 minute off-peak headway.

During peak hours, extras on both routes departed from the depot and from the sidings at Southbank Road and Binns Road which catered for employees at Crawfords, Meccano, Littlewoods, Automatic Telephone and Paton Calvert. Inbound short workings included the depot, Durning Road, Crown Street and Clayton Square and outbound the depot, Edge Lane roundabout, Springfield Road crossover (very occasionally), Broadgreen (6), Pilch Lane (40) and Finch Lane (40). Most extras just displayed the ultimate destination on the screen, sometimes without a route number. Specials also provided at the end of visiting hours at Broadgreen Hospital working inbound to Durning Road, Crown Street or Clayton Square.

"My friend Chris Bennett and I boarded 266 and left the Pier Head at 8.02am bound for St Oswald's Street. During the journey, I persuade the reluctant driver to reverse the car on the rarely used Springfield Road crossover rather than circling the roundabout. As a result, this becomes the last car to use this crossover. Watching 266 depart for the depot, we board 214 bound for Durning Road – the last car to use another crossover. We then take 235 and 296 to complete the journey to Bowring Park. It is by now about 9.30 and the crowds of young children at the terminus make photography extremely difficult. My friend and I wait for cream car 293 to take us to the depot where we watch early morning peak hour duties running in, many for the last time. Boarding 263 we head towards Pagemoss, but on the way down Brookside Avenue, we pass disabled 272 stranded on the inward track. Alighting we eventually witness the spectacle of 263, back from Pagemoss, attempting to push 272, with little success, until 257 also arrives and between them, accompanied by much sparking and flashing, they manage to get 272 on the move. We heard later that 272 had track brake trouble." **(MJ)**

E C Bennett and Martin Jenkins/Online Transport Archive

The Corporation issued the following instructions to be implemented during the final afternoon: "The replacing buses will in all cases take over from trams at the outer terminals, and will observe the tram routes and stops." This evocative view shows a driver handing his time board to the replacing bus driver at Bowring Park. Trams returning to the depot remained in service whilst, for the rest of the day, buses maintained the tram schedules. *[Above] Brian Yates collection*

ROUTE PROFILE
6, 6A, 40

Leaving South Loop, Pier Head. *J B C McCann/Online Transport Archive*

Lord Street. *Peter Mitchell*

Clayton Square. *J W Gahan collection*

From the top of Elliot Street, the 6s reached the junction of Crown Street and West Derby Street via Lime Street and Pembroke Place (seen here). *Peter Mitchell*

From the top of Elliot Street, outbound 40s reached West Derby Street by way of Brownlow Hill (seen here) and Crown Street. *Peter Mitchell*

Paddington. Inbound cars on all three routes descended Paddington, scene of a fatal runaway in 1934, with the 6s turning into Crown Street and the 40s onto Brownlow Hill. *A F Gahan/Online Transport Archive*

Crown Street. The remaining track was used by inbound 6s but outbound 40s. It had not been relaid for years whilst the other track was relaid and repaved literally weeks before it was abandoned! *H B Priestley/National Tramway Museum*

Outbound 6/6A/40 climbed up to Edge Lane firstly via West Derby Street. *N N Forbes/National Tramway Museum*

218 | THE LEAVING OF LIVERPOOL

The outbound climb continued up Mount Vernon and North View (seen here). *Peter Mitchell*

The inbound one-way workings began along Towerlands Street. *Peter Mitchell*

Citybound peak hour cars outside Edge Lane depot. *H B Priestley/Online Transport Archive*

Edge Lane, Binns Road peak hour siding. *Pam Eaton/National Tramway Museum*

Approaching Edge Lane roundabout. *Peter Mitchell*

1957 | 219

ROUTE PROFILE
6, 6A, 40 *(continued)*

Edge Lane/Mill Lane. In the background is the start of the city's first section of reserved track, opened in 1914. *N N Forbes/National Tramway Museum*

At the Gardeners Arms, the 40s continued along Edge Lane Drive in the direction of Page Moss. This view is taken near Broadgreen Hospital. *Peter Mitchell*

Thomas Lane, Knotty Ash, passing the home of Ken Dodd. *Peter Mitchell*

Turning from East Prescot Road onto Brookside Avenue. *J B C McCann/Online Transport Archive*

East Prescot Road, Finch Lane. *J S Laker*

220 | THE LEAVING OF LIVERPOOL

From the Gardners Arms junction, the 6s continued east towards Bowring Park. *Peter Mitchell*

Broad Green station. *R B Parr/National Tramway Museum*

Broad Green Road, crossing the Cheshire Lines Hunts Cross-Aintree railway line. *F N T Lloyd-Jones/Online Transport Archive*

Journey time	30 minutes (6), 34 minutes (6A), 36 minutes (40)
Last day of operation	14 September 1957
Cars seen in service on the last day	207, 213, 214, 224, 226, 231, 235, 237, 238, 245, 252, 254, 257, 258, 260, 263, 264, 266, 271, 272, 274, 280, 288, 293, 296
Last cars	Springfield Road crossover (266), Broadgreen Hospital special (226), Durning Road (214), Crown Street (258), Clayton Square (237), Broadgreen (207). Last 6 (207), Last 6A (293), Last 40 (274)
Track abandoned	All remaining track

LIVERPOOL CORPORATION PASSENGER TRANSPORT

TRAM ROUTES

6, 6A, 40

CONVERSION TO BUSES

REVISION OF SERVICES TO

HUYTON, WEST DERBY AND BROADGREEN

NEW LIMITED STOP SERVICE

506

HUYTON—CITY

COMMENCING SUNDAY, 15th SEPTEMBER, 1957

24 Hatton Garden, Liverpool 3. P. 2180

'Phone CENtral 7411

W. M. Hall, General Manager

BOWRING PARK—CITY

Tram Route 6, 6a Stage No.							Stage No.
18	Bowring Park (Childwall Lane)	...	...	...	...	...	13
20	1½	Bowland Avenue	...	...	...	...	11
22	2½	1½	Broadgreen Road (Edge Lane Drive)	...	...	...	9
24	4	2½	1½	Church Road	...	...	7
26	4	2½	1½	Deane Road	...	...	5
28	5	4	2½	1½	Edge Hill Church	...	3
30	5	4	2½	1½	Lime Street	...	1
—	6	5	4	2½	1½	City	

"Crammed to capacity and watched by a few curious citizens, the last 6A leaves the Pier Head at 1.30pm. After photographs at Bowring Park it eventually reaches Edge Lane just before 2.30pm and is hastily taken away in order to be made ready for the final procession." (MJ)

J B C McCann/Online Transport Archive

"The following 40s leave the Pier Head – 213 at 2.36pm driven by Harry Tindale, 264 at 2.39pm and 263 at 3pm. These are followed by 237 which comes into town as far as Clayton Square where it departs at 3.10pm bound for the depot. Now for the final 40. I queue at Pier Head for over an hour and just manage to get an upstairs seat on 274 the very last car in ordinary passenger service which leaves at 4.02 pm driven by H Evans accompanied by two inspectors. Virtually non-stop run – no one able to get on and no one eager to get off. Everybody talking – loudly. I want to just hear the motors and soak up the atmosphere. 274 attracts considerable attention, possibly due to chalked slogans on front and rear dashes 'Last 40' etc. Cameras click and whirr at every corner and little knots of people stand and wave as we rumble by. 36 minutes later we reach Pagemoss. Everyone turfed off. More photos. Again just managed to get back on. My last journey on a Liverpool tram. Unreal. Again too much loud talking – even laughter coupled with the roar of the corrugation on Thomas Lane. Suddenly we're there. Outside the depot – large crowds. At just after 4.52pm, I walk down the stairs onto the back platform and into the road. Ask driver H Evans to sign my ticket. Watch as 274 edges cautiously into the depot. Quite tight security. For me that was it." (MJ)

J B C McCann/Online Transport Archive

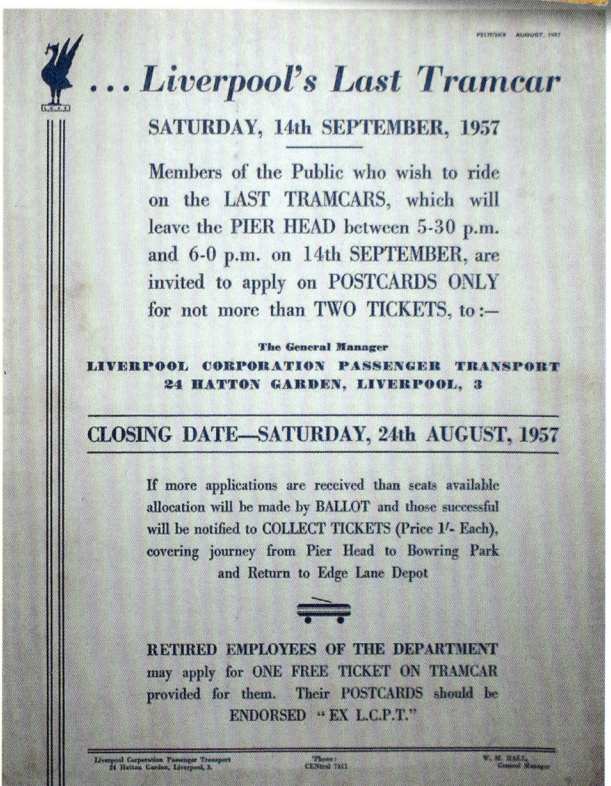

"Tickets were assigned by ballot and I was unlucky. At least I could take some photos. Managed to team up with someone with a car so by back street dodging we were able to see the procession at different locations." (MJ)

Cars forming the last procession (letter in parade and driver where known): 210 (decorated; N, 'Nobby'), 264 (M), 214 (L, Walter Benstead), 235 (K, 'Athlete'), 213 (J), 226 (H, Jim Minnery), 266 (G, Fritz Pullman), 260 (F, Harry Tindale), 296 (E), 252 (D), 245 (C, carrying retired staff, 'Rastus'), 207 (B, also carrying retired staff), 293 (A, carrying civic dignitaries, driver Tom Webster with G Rodden acting as conductor).

The Transport Department arranged the event with military precision issuing very precise instructions. "13 trams will leave Edge Lane depot from 16.30 hrs, in order N. M. L. etc to A. They will travel out of service to Pier Head via Route 40 with rear doors closed. Indicators will display 'Private'. All Drivers and Conductors will wear full serge uniform, including caps, and as neatly as possible. Lettered cards will be displayed in rear nearside window. Conductors will collect parcel of brochures from D.I. before departure."

235 (K), which has just passed beneath the closed Overhead Railway, is followed by 226 (H) and 266 (G) to take up position for the procession. *[Opposite, right - centre] Peter Mitchell*

222 | THE LEAVING OF LIVERPOOL

Liverpool's Last Tram

SATURDAY SEPTEMBER 14th

1897–1957

THE LAST TRAM WILL LEAVE PIER HEAD AT 5·55 p.m. TRAVELLING VIA LORD ST., CHURCH ST., LIME ST., LONDON RD., PEMBROKE PL., EDGE LA., BOWRING PARK RD., TO BOWRING PARK AND THEN RETURN TO THE EDGE LANE DEPOT AT APPROXIMATELY 6·45 p.m.
- INTERESTING ILLUSTRATED HISTORY ON SALE — PRICE 6d.

LIVERPOOL'S LAST TRAMCAR

The Bearer is allocated a place on Tramcar

L

(No. 214)

Leaving PIER HEAD, 5.30 p.m., on SATURDAY, 14th SEPTEMBER, 1957

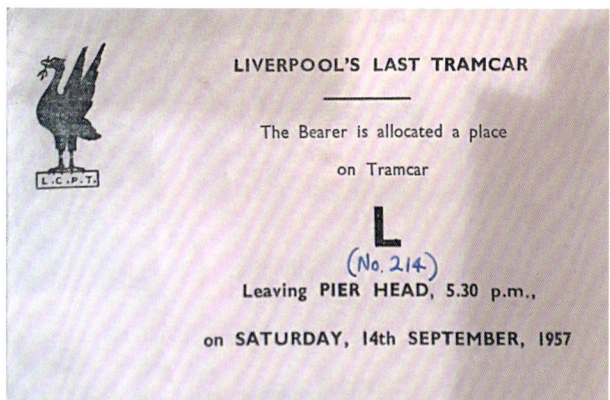

"On arrival at Pier Head trams will park to instructions on the South Loop. (Trams C. B. A. and possibly D. on spare line). At Pier Head, indicators will be set to '6A – Bowring Park – Private'. Drivers and Conductors will stay with trams – if they must be relieved an Inspector will be asked to take over. Rear doors will be opened and ticket holders permitted to board on production of ticket letter in accordance with tram letter. Pink tickets (D to N) will be collected. White (A), Yellow (B) and Blue (C) tickets must be shown but retained by passengers." *[Right] Leo Quinn collection/ Online Transport Archive*

"At 17.45 trams will move off in CLOSE CONVOY at a moderate speed, preceded by a Police Pilot car. Traffic lights will probably be switched in favour of the convoy and point constables will afford clear passage. A police motor cycle escort will accompany the last tram (A). An Inspector will be stationed between Mill Lane and Sturdee Road to open intervals between trams and an Inspector at Bowring Park will open these intervals

still further. Once the journey has started Conductors will issue a brochure to each passenger (one each included for Driver and Conductor)." In the first view, the procession cars are bunched as they parade along Lime Street headed by car G (266). A few minutes later, Alun Owen's famous play *No Trams to Lime Street* would be all too true. Decorated 210 led the procession along Edge Lane. In the second view, 252 is preceded by 296, 260, 266, 226, 213 and 214 as they pass the depot in convoy. *[Top, above]*
Leo Quinn collection/Online Transport Archive; Martin Jenkins/Online Transport Archive

"At Bowring Park trolleys will be turned by a Regulator. 'Edge Lane' will then replace 'Bowring Park' on indicators." 245 is surrounded by crowds at Bowring Park. *[Above, right] Peter Mitchell*

"On arrival at Southbank Road stop trams D to N will off-load as quickly as possible (Conductors will ensure nobody remains on) rear door will be closed and trams proceed directly into the Depot with trailing trolleys. Trams A. B. & C. will not unload but will proceed directly into the Works covered way and unload near to the traverser. The Band will play at the Pier Head from 16.45

hrs until convoy departs, then travel by bus to Edge Lane, playing there until the last tram enters the Works". Complete with its police escort, 293 approaches Edge Lane depot for the last time. The band will be playing in the background. *[Opposite, top] Peter Mitchell*

"Everything nearly went according to plan. Dense crowds had gathered. As the procession left, the Liver clock was striking 6pm and a great cacophony of whistles and hoots came from ships in the river. A stirring tribute. Substantial crowds especially at Bowring Park and around the depot. Bus drivers tooting their horns as they pass. Parents holding up their children. Scores of pennies laid on the rails. As directed, most cars did show 6A but some on board turned the blinds to numbers such as 19B and 29A. Some long-serving drivers clearly upset. I felt Corporation just wanted the trams gone and forgotten. By 7pm it was over and the crowds had dispersed." **(MJ)**

224 | THE LEAVING OF LIVERPOOL

"There was no disorder, no souvenir hunting, and no fuss. Never in recent years has a last tram ceremony passed so decorously" *(Modern Tramway)*

"No souvenir hunting – not quite true. For me the day was not yet over. I had long wanted a tram stop to keep as a souvenir. Along with a couple of other likeminded people, as night fell we were on the roof of a van attempting to remove a blue request stop. Sensing something suspicious a passing police car pulled up and the copper asked what we were doing. Better tell the truth – so we explained. 'Well you won't get it off with that' was the response as he opened the boot of his car and produced a large spanner. Within minutes the 'theft' was completed and he was on his way. Today it hangs in my kitchen." **(MJ)**

Edge Lane depot and works: By the next morning the depot (F on the plan) was cleared of trams and work began on converting it for use by buses, these having hitherto operated from a building at the back of the site, referred to as 'the Back Shed'. The depot had two narrow entrances with the tracks splaying out inside, although for some time the rear entrance had not been in regular use. When trams returned to depot, the trolley was turned before the car entered with the conductor monitoring progress through the frogs. As a result, the poles

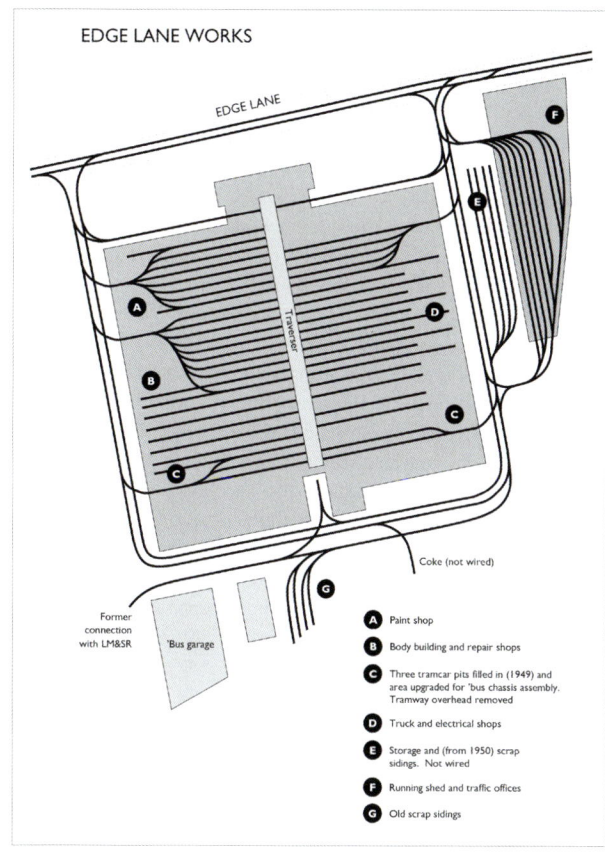

EDGE LANE WORKS

- **A** Paint shop
- **B** Body building and repair shops
- **C** Three tramcar pits filled in (1949) and area upgraded for 'bus chassis assembly. Tramway overhead removed
- **D** Truck and electrical shops
- **E** Storage and (from 1950) scrap sidings. Not wired
- **F** Running shed and traffic offices
- **G** Old scrap sidings

were always in the correct position ready for the next turn of duty. The conversion work consisted of removing tracks and cleaning galleries and laying a new floor, but otherwise the building remained virtually unchanged throughout its existence. It was last used operationally by buses in 1991, although it was utilised for a variety of related purposes for several years thereafter.

Although limited tram movements continued to occur within and around the works into the first part of January 1958, it now only handled buses. When opened in 1928, it was one of the largest, purpose-built works in the country with a capacity to build 1000 trams. If the war had not intervened, would it have produced more Baby Grands using some recycled equipment, or would it have been reconfigured to construct American-style single-deckers? In the event it did build hundreds of new trams although the workmanship may not have been to the highest standard. The reduced workforce also kept the increasingly neglected fleet on the road during the dark days of the war. This diagram shows the layout at its maximum. Over the years, areas were converted for buses and unwired scrap sidings were laid. After tramway abandonment, it continued repairing buses, latterly including outside contract work, into the 1990s. The entire site, including the fine office building with its imposing clock tower, was cleared in early 1997.

15 September 1957: When the depot was cleared of trams, the survivors were taken under their own power for storage in the Works until there was space on the dump. The Corporation earmarked 245 for preservation whilst 293 was acquired by an American museum for the scrap value of £250. Attempts to sell the 1948 railgrinder failed and the car was broken up in mid-January 1958. *[Above] R L Wilson/Online Transport Archive*

10 October 1957: LRTL representative, Ted Gahan, received a letter from the General Manager informing him that "the Transport Committee agreed to offer a tramcar to the New England Electric Railway Historical Society" and "the Committee had agreed to retain a second tramcar for preservation by the British Transport Commission or any other suitable Transport Museum in the country."

11 October 1957: "245 and 293 ran in Edge Lane from Littlewoods to the middle of the Works on outward line." (AFG) No known photographs.

CHAPTER 15 | 1958 and beyond

Although the trams finished in 1957, the story doesn't end there. For several decades, many reminders of the system could still be seen in the city, if you knew where to look, and some still remain. Through the dedicated work of preservationists, it is even possible to ride on genuine Liverpool on the Wirral Heritage Tramway in Birkenhead and at the National Tramway Museum at Crich in the 21st century. The full story would justify a book of its own, but we have endeavoured to give an outline of the last 60 years or so here.

15 January 1958: Scrapping nearly completed. Included was SP1 the last classic Standard car.

18 January 1958: 271 was the last passenger car to be broken up. *[Below] J B C McCann/Online Transport Archive*

27 January 1958: Appropriately, the last car to be scrapped was also the oldest. This was CE&S 283, which had started life as an open topper in 1901. *[Right, top] J B C McCann/Online Transport Archive*

7 May 1958: 293 departed for the United States and the Seashore Trolley Museum in Maine. It was moved to Gladstone Dock on a transporter trailer hauled by Pickford's ex-Army Scammell Pioneer GKD56. The combination had to travel via Ford, Netherton, Crosby, Waterloo and Seaforth in order to avoid low bridges. *[Right] J and C Moores/National Tramway Museum collection*

At Gladstone Dock, the car was hoisted onto the United States Lines *American Packer*. On arrival in the USA, 293 spent many years out of doors even during the bitter winter months. As a result its body has suffered. For years, there has been talk of possible repatriation. *[Overleaf, top left] Martin Jenkins/Online Transport Archive*

1958 and beyond | 227

Work progressed on dismantling the city's tramway infrastructure. Much of the street track was simply tarmacked over but, on the reserved sections, most was lifted and sold for scrap. The former 'grass tracks' were either retained as wide central reservations, or absorbed into carriageway widening schemes. The section of Roby Road near the former terminus of the 6A is seen here during 1958. *[Below] J B C McCann/Online Transport Archive*

8 June 1960: 1055 became the last Liverpool tram to run on a public street when, in the early morning, it was moved from Dalmarnock depot to Coplawhill Works. From there, it was taken to the site of a proposed museum near the Middleton Railway at Parkside, Leeds, the transportation cost of £130 being paid for by Guinness Ltd. It is pictured being unloaded in Leeds on 9 June 1960. *[Above] Frank Atherley/Online Transport Archive*

9 June 1958: First Liner withdrawn in Glasgow, No 1024, formerly Liverpool 927, the first to arrive in 1953. Others followed rapidly until 1036 (891) is the last to go in July 1960.

November 1958: The Secretary of the newly formed Liverpool University Public Transport Society (LUPTS) wrote to Liverpool Corporation with a tentative plan to operate 245 on part of the recently-closed West Kirby-Hooton railway line. The request was rebuffed: "… regarding Tramcar No.254 (sic), … the Committee did not feel that they could turn this car over to your society … due to the strong body of opinion which considered that the tramcar should be retained in Liverpool. As you know we are hoping that arrangements may be made with the Director of Museums in Liverpool for this tram to be taken over by him but meanwhile it is intended to retain it in our Edge Lane Works." LUPTS turned its attention to preserving one of the Liners still operating in Glasgow.

21 February 1960: To help raise funds to purchase 1055 (869), LUPTS organised a six-hour marathon tour in Glasgow. Eventually the car was acquired for £50 by the Merseyside Tramway Preservation Society (MTPS). *[Right, top] Frank Atherley/Online Transport Archive*

"I shall never forget the early morning move. After photographs at Dalmarnock, securing some tape recordings for Frank Oldfield en route, I then drove the car through the city centre and over Glasgow Bridge to just before the Works. So other than the driver, I was the last person to operate a Liner on public streets." **(MJ)**

May/June 1961: The Liverpool sound was finally silenced in Glasgow when 1954 Coronations 1394 and 1398, which had been built with trucks and motors salvaged from the Green Lane depot fire in 1947, were given Glasgow bogies and motors from withdrawn Coronations.

October 1961: Owing to vandalism in Leeds, the MTPS made the decision to move the car from Parkside to Crich. An announcement was made at the LRTL meeting in Liverpool on 12 October 1961 that the car would have to be moved within a month, at a cost of £100. "A collection was made at the meeting which realised the fantastic sum of £23/17/6!" (AFG) This decision saved the tram as other cars stored alongside it, including the only surviving ex-Swansea and Mumbles car, were vandalised beyond repair and ultimately scrapped.

26 November 1961: 869 was moved to open but secure storage at the Crich tramway museum in Derbyshire.

21 and 28 January 1962: First of many working parties on 869 in which Tony Gahan participated. *[Below] Harry Haddrill/Online Transport Archive*

"Unfortunately it was lashing with rain so we could do very little exterior work. We put in a couple of windows and made the other broken ones more or less waterproof. I spent most of the time taking out the Glasgow lighting frames which had been securely fixed at Coplawhill Works. We cleaned out the piled-up platform and I fixed all four doors, putting them back on their tracks and oiling them. Brian [Martin] painted the front panel and put on the numbers 869. I also cleaned odds and ends and generally weighed up what has to be done on future visits. It really is a tremendous job we are undertaking as she is not in a particularly good shape."
(AFG)

11/12 February 1967: Over a two-day period, 869 was moved to Green Lane depot for further restoration to take place. It was photographed en route passing through Holmes Chapel in Cheshire. *[Right, top] J G Parkinson/Online Transport Archive; Brian Martin/Merseyside Tramway Preservation Society*

19-21 July 1973: After many years stored in the paint shop at Edge Lane, 245 was cosmetically restored and displayed at the Liverpool Show held at Wavertree, after which it returned to Edge Lane. *[Right, centre] Brian Martin/Merseyside Tramway Preservation Society*

28 October 1977: Having been in use as a bowling green clubhouse since 1955, the lower deck body of 762 was removed from Newsham Park for preservation. Restoration work was undertaken at Green Lane, Speke, Princes Dock and in a covered space at Cammell Laird's in Birkenhead. *[Right, bottom] Brian Martin/Merseyside Tramway Preservation Society*

9 January 1978: 245 loaned to the Steamport Transport Museum at Southport.

1958 and beyond | 229

29 September 1979: 869 displayed at the Big Bus Show at Edge Lane on a lowloader before moving back to the National Tramway Museum at Crich. *[Above] Brian Martin/Merseyside Tramway Preservation Society*

25 April 1988: 245 transported from Steamport to the Merseyside Museums' Large Objects Collection at Princes Dock, Liverpool. *[Below] Brian Martin/Merseyside Tramway Preservation Society*

Autumn 1990: The Large Objects Collection closed to visitors and 245 no longer accessible to the public. Other historic trams were moved into temporary storage by the MTPS.

20 December 1992: Now fully restored to working order, 869 first ran under its own power at Crich. *[Right, centre] Charles Roberts/Online Transport Archive*

14 April 1995: The Wirral Tramway heritage line opened at Birkenhead and 762 is moved to its Shore Road premises for continuing restoration.

2 October 2000: 762 is displayed at the Wirral Transport Museum's new premises in Taylor Street, fully restored. It entered public service in 2001. *[Above] Charles Roberts/Online Transport Archive*

2001: Merseytram project proposed a three-line modern tramway network for Liverpool.

230 | THE LEAVING OF LIVERPOOL

November 2005: Liverpool One transport interchange opened. Tramway warning signs are installed but quickly covered up.

29 November 2005: Work on Merseytram cancelled due to concerns over costs.

May 2006: 245 is moved to the Wirral Transport Museum for restoration. Work was undertaken by the MTPS, partly funded by a National Lottery heritage grant.

October 2013: Merseytram project formally closed.

12 September 2015: With its restoration complete, 245 entered service in Birkenhead. *[Above] Charles Roberts/Online Transport Archive*

In 2020:
Fully restored, 869 is now a prize exhibit at the National Tramway Museum in Derbyshire.

Following years of neglect and long-term storage, 245 has been magnificently restored by the MTPS and operates on the Wirral Heritage Tramway along with Liverpool 762, horse car 43 and other Merseyside area trams.

Although undercover, 293 is non-operable at the Seashore Trolley Museum. *[Right] T P Kane*

1958 and beyond | 231

[1]

[2]

[3]

[4]

[5]

[6]

Virtually nothing survives of Liverpool's once great tramway but there are still miles of the former central reservations, some traction poles and even a few rosettes which once supported the overhead wires. Some very short sections of tram track are visible whilst others, long buried under tarmac when routes were abandoned, emerge during roadworks.

From an earlier era, a short section of horse tram track has been preserved in Franklin Place, Anfield, the site of a depot which closed in 1900. [1]

Until about 1936, specially converted single-decker trams, known as 'money cars' made deliveries of cash and, in some cases, stores into two short stub sidings located in the Hatton Garden Headquarters Building. These survive today and can easily be seen when the goods access doors are open. [2]

Over 20 rosettes survive in various parts of the city, with building owners probably oblivious to their former use. The location of this one in Fazakerley is clear. The wiring it supported was last used in November 1951. [3]

The tracks for Napier siding at Gillmoss, complete with crossover, were still visible in September 2006, 50 years after they were last used, but have since been removed. Ironically, the alignment would have been used by the failed Merseytram project. [4]

A small number of tram poles survive, but this complete set on Utting Avenue East was uprooted in June/July 2019. The provision of space for the reserved tracks is evident. [5]

Surviving for 64 years from the closure of the Kirkby routes, one of the most significant set of remains were the tracks embedded in the tram-only bridge over Knowsley Brook at Radshaw Nook. With no regard for their heritage value, they were unceremoniously buried under a layer of tarmac in September/October 2020, perhaps one day to be rediscovered by a new generation of tramway enthusiasts. [6]

All Charles Roberts/Online Transport Archive

Other tangible survivors:
Special work from Edge Lane Works at Crich
Liner seats acquired from Glasgow and installed in a coach on the Ashover Light Railway
Controller and motor used at the National Waterways Museum at Gloucester Docks
Two Dick Kerr K3B controllers on Bolton 66
Trucks and motors in Leeds 600
Front panel off 181 and a Liner seat at Wirral Heritage Tramway
Memorabilia ranging from stop signs to fleet numbers in public and private collections
Wealth of photographs, colour slides, films, time-tables, leaflets etc in public and private collections

Do you know of anything else?